Contents

Foreword

Let me start with a true life story. When, in 1982, I proudly brought home our first 'microcomputer', I responded to the asked 'What is it for?' question of my inquisitive 8-year-old son, Basil, by saying that it was a clever machine, which we can teach and feed with information, and which it can recall when asked. I connected the machine and completed the installation of the unforgettable Disk Operating System and the word processor, and tested these. As a 'demonstration', to a solemnly watchful Basil, of the 'cleverness' of that machine, I typed a couple of lines into the word processor, making errors and correcting these, and saved these as a document onto a diskette and switched off. Then I switched on the machine and recalled the document in which the two lines were saved. Watching with what I thought was awe, Basil silently and slowly nodded his head with what I thought was his concurrence that it was indeed a clever machine. He remained glued to me and the machine until I had to get up and do something else – before which I opened a blank document and encouraged him to enter whatever he wanted. When I re-appeared, Basil immediately announced, pointing to the screen, that it was not a clever machine at all! I looked closely and found that he had patiently and correctly typed in, '*My name is Basil. What is my name?*'. I looked at him and he daringly remarked 'It can't even recall a name.' And he sat there waiting for a response which was certainly not to come! He did not even 'press <enter>', which even if he had, would not have come anywhere near an 8-year-old's spontaneous view of what a 'clever' machine could be capable of.

That incident taught me two lessons. First, a profound professional alertness on how we should explain, introduce and teach computer-assisted anything. Second, the significance of the human–machine interface. As that incident flashed in my mind again and again over the years to come, I realised that I did not fathom the depth of an 8-year-old's observation in terms of what it really was: **both** as a real-life spontaneous expression of the difference between knowledge and mere collections of documents, **and** a signal of the potential of 'computing' in that realm.

Since then, computing and those 'un-clever' microcomputers have come a long way. Not only are we learning and developing more and novel uses of computers, but the very paradigm of computing has significantly evolved. The health sector is a fine witness to that. We, in the health sector, initially mimicked, indeed literally copied, some of the business sector's *applications* of computing in the clerical, administrative and financial functions. Then we developed computerised health information systems, models and decision-support systems, and we entitled the field, '**medical informatics**'.

Two other major milestones in computing technology brought about significant new uses of informatics in the health sector. These are multimedia and computer-assisted telecommunications, the latter earning the broader title of '**health telematics**'. Since then, informatics and telematics have gone into many aspects of health and medical care: *from* being implanted as microchips in, and monitoring and supporting the functioning of, parts of the human body *to* relaxing the time honoured face-to-face encounter between a patient and a care giver. Indeed, the terms medical informatics and health telematics gradually grew into '**TeleMedicine**' and '**TeleHealth**' and then, almost with the recent turn of the millennium, into the even broader '**eHealth**'.

These are a myriad of terms when viewing 30 or so years against the age of humanity. It is also a reflection of both the developments in the power and affordability of the technologies involved, and the growth of relevant applications of computing, networking and telecommunications support to the health sector including medical care. I carefully choose to express it as 'relevance' of such support, rather than as actual, outright 'uses', for a good reason: to reflect the enormity of the jump from proving and demonstrating the effectiveness of a methodology and a technology, to an operational, cost-effective, safe and ethical eHealth – in other words to actual, regular uses in real-life healthcare settings.

Healthcare is not banking or insurance or library services, which involve transactions whereby the locations and identity of whoever one deals with do not matter – clear cut transactions with clearly set rules and regulations. A health, particularly medical, care transaction is a highly personalised one-to-one service, with widely differing data and means of communication. Whereas many aspects of eHealth can indeed be governed by the eCommerce agreements, it would be a mistake to believe that these eCommerce agreements would suffice for eHealth. More needs to be learned, and assessed, on the continuum and whole span of eHealth. In the meantime, experience shows that TeleMedicine remains the eHealth area that captures the attention of patients, healthcare professionals, policy makers and politicians, as a means of directly improving equity of access to quality healthcare.

My WHO duties involved, and do involve, direct support to and collaboration with member countries in eHealth – mostly the introduction and uses of informatics and telematics in health, particularly TeleMedicine, in both the industrially developed and developing countries. Such collaboration with countries is a technical challenge, a moral load, a pleasure and a privilege – to collaborate with countries to analyse and elaborate their requirements, to work out and articulate the justification for solutions including e-solutions and their potential cost-effectiveness, to technically specify the solutions sought, and to assist in comparative analysis of the solutions offered. The true strength of our role stems from the privilege and the opportunities to witness and participate in such efforts in many countries, thus enabling us to assemble a truly *global view* of eHealth developments in a great variety of cultural, socio-economic and political settings, as well as of differing visions and infra-structure. That privilege and those opportunities are behind the views expressed in this Foreword.

To me, one of the most surprising, present, global realities is the *high level of awareness* of TeleHealth and TeleMedicine by the politicians and policy makers in the great majority of countries, including the poorest and the least industrially developed countries. Most readily cite what TeleHealth and TeleMedicine is, why they require to introduce these and the impediments to their doing so. My optimism for healthcare in the industrially developing countries is renewed every time I encounter that striking reality – and recently that had been frequent and rapidly increasing.

My optimism does not contrast, as it may initially appear to, with two recent statements from two totally differing, internationally well-recognised institutions. The first statement is from the Dhaka, Bangladesh, based *Bytes for All* whose charter includes: to expose the 'inequities' caused by the 'digital divide' and to educate people worldwide on its impact and identify ways that the divide can be closed. *Bytes for All* points out 'Some of the daunting realities of the third world: a modem costs more than a cow', 'access to telecommunications is a privilege' and 'information by definition is a property of the government'. The second statement is from the BBC which in a recent report states:

> 'The hype for everything online obscures the reality about how technology is changing life at the end of the 20th century. From Manhattan to Madrid, the Internet has fundamentally changed work, recreation – even love. But in Malawi and Mozambique, life remains very much the same. More than 80% of people in the world have never heard a dial tone, let alone sent an email or

*downloaded information from the World Wide Web. "Think how powerful the Internet is. Then remind yourself that fewer than 2% of people are actually connected", said Larry Irving, former US Assistant Secretary of Commerce. "The power of the Web increases exponentially with every person who goes online. **Imagine what we're missing**."'*

In fact, Mozambique is one of the few African countries that has an operational TeleMedicine (Radiology) service. But the BBC report's point is well made, and I fully share Larry Irving's concern.

There is ample justification for, and we urge, a Marshall Plan-like global effort to enable the other 80% of humanity to contribute to, and benefit from, the Web challenges and what it offers and stands for.

It is widely recognised that eHealth differs from the classical model of care giving in that, because it is largely Internet based, it works with anyone, wherever he/she may be and at any time. Yet, our experience points clearly that *TeleMedicine is required predominantly within nations and to a much lesser extent between nations*. Thus, whereas TeleMedicine is leading towards a 'global clinic', it is justified and required first and foremost to provide national links and services. For eHealth, however, the 'global' dimension is vital and much needed, particularly for training, education and technology transfer. But it will take longer to evolve depending on the speed, quality of, and consensus on the requisite secure international authentication, certification and accreditation procedures.

The advent of eHealth, and particularly TeleMedicine, appears to place the health sector at a major crossroad represented by two differing views. The first view stipulates that eHealth is but a powerful technological support to the traditional care-giving model which will basically remain unchanged by the eHealth technology except for improvements in efficiency, efficacy and economies. The second view stipulates that eHealth is the trigger and the actual start of a totally new healthcare paradigm, whereby the very relationship between the individual, the care provider and the health authorities will be completely altered. I believe that, ultimately, the second view – that is, the emergence of a significantly altered healthcare paradigm – will prevail but after a period in which both views will co-habit, with the first view gradually converging onto the second view. But the new healthcare paradigm calls for major national and international thinking and understanding on, for example, the changes of the patient–doctor relationship; the changes in public or professional behaviours; the emergence of inter-dependent but geographically dispersed providers of care in place of relatively 'local' independent providers; and the vetting, certification and clinical audit of the remote services.

eHealth, and particularly TeleMedicine, has hitherto been driven by the unavoidable attitude towards the novel: excitement, enthusiastic experiments and pilot installations, mostly on point-to-point links, and rushed introduction in a few institutions, leading at times to staff resistance. But experience is showing that the depth and potential of eHealth, particularly TeleMedicine, are such that if TeleMedicine is indeed cost-effective, publicly accepted and is to be regularly used, it must be formally recognised and integrated into the healthcare services. As such an *eHealth policy and strategy* is needed that is an integral part of the overall national health policy and strategy. This is no simple task and, even with hard evidence of its benefits, would require that policy decision makers be politically persuaded to give eHealth a chance with the other imperatives because, invariably, its resource requirements have to be met by re-allocation of the limited budgets. Such policy and strategy should also include *inter alia* plans for training of the health and medical practitioners; for introducing new care protocols; for record keeping of medical transactions, and storage, access and controls of such records whether they are kept by a care provider, by the individual or distributed amongst care and service providers and virtually assembled when needed; set rules and fees for specific TeleConsulations; and rules for fair sharing of revenues between the participating institutions and individuals.

All essential parts of healthcare are regulated and quality assured (e.g. pharmaceuticals, measuring instruments) for ethico-legal reasons, mainly aimed at protecting patients and society at large, guarding against avoidable risks and ensuring fair commercial practices. eHealth, and particularly TeleMedicine, is no exception and must be *quality assured* and *regulated*. Some issues are relatively straightforward to deal with in that the existing rules and regulations – for example, Data Protection – would be extended to cover 'remote' eHealth transactions. Other issues stem from the truly novel uses of eHealth and call for original analysis, thinking and resolution, e.g. the use of measuring and monitoring devices which are implanted in or carried on a patient's body, and equipped with transmitters to enable the TeleSurveillance of the patient by care providers. Under what circumstances could such TeleSurveillance circumvent the 'privacy and confidentiality' of the patient? And when is such a breach a lesser evil than the medical problems it seeks to monitor or overcome? All these issues are further complicated by the fact that eHealth also involves e-images and video clips, which calls not only for the protection of the patient but also of any and all individuals appearing in these images including care providers and family members.

The ethico-legal issues in eHealth call for another frank remark related to *e-records*. It is a fact that these are much more secure than paper records (just visit a paper records department in a local hospital or clinic, or trace a borrowed record!). The concerns over the security and confidentiality of patient-identifiable data in e-records are legitimate and stem in part from a wide recognition that the damage of a breach of e-records, when it happens, could be dramatically larger, in scope and geography, than a security breach of paper records. These are further complicated by another fact – whereas encryption techniques could eliminate the risks of a breach of e-kept patient-identifiable data, the laws of some countries (including eHealth champions) inhibit extensive encryption of such records.

TeleEducation is intimately related to TeleMedicine, both in content and in technological support. Indeed, all the country TeleMedicine projects for which I was and am personally responsible have a significant education or training component. Indeed, the justification for some projects was finally accepted because of the heavy component of TeleEducation, mostly for Continuing Medical Education. But the issue is deeper than that. Even ordinary TeleMedicine practice has a learning/training impact in that repeated TeleConsultations lead to skill acquisition. What skills? And are these compatible to the norms and standards approved for the country in question? There are thus strong calls and arguments for a peer review and/or a clinical audit of such remotely provided services.

The most popular and recognised uses of eHealth, particularly TeleMedicine, are those that improve the quality, efficiency, efficacy and economy of providing healthcare services. There is a *shortage of hard evidence* of the merits of eHealth. A great deal of effort should go into, and is going into, assembling and presenting such evidence. Any such evidence would be, as per the original justification, compared to a denominator of existing services. Yet, eHealth and particularly TeleMedicine, in some parts of certain countries such as South Africa, is the only means by which any healthcare has ever been provided. That should be enough hard real evidence.

Another burning message: God protect the health sector from the over-simplifiers who assert that 'technology is not a problem'! It is true that gone is the era when the problems of incompatible operating systems, of incompatible application packages and of proprietary standards were viewed as issues for the client to resolve! It is true that the power of the Internet, and its non-commercial and non-vendor origins, led to widely and readily accepted transmission protocol, hypertext mark-up language, extensible mark-up language, and the friendly browser technology. It is true that these led to *de facto*, international and widely available tools for any one in the world able to connect to, draw from or contribute to the Internet. These are great improvements in technology. But the technology remains a problem for a truly global eHealth – relatively high costs particularly of peripheral devices, incompatibility of peripheral devices, demanding power consumption, maintenance, etc. These beg for dramatic improvements, for a step-jump of improvements, before technology is viewed as a non-problem!

About 75% of current TeleMedicine is supported by store-and-forward, offline communications. These facilities cannot support the full eHealth which would require multimedia (broadband) communication services. And the health sector in most countries may not master the resources needed to build the key parts of the national infrastructure to fully support eHealth. A *multi-sectoral approach* is then in order, whereby the health sector shares the required networking and communications services with other sectors, such as education, libraries, interior or private enterprise. For the remote, small villages in the least industrially developed countries, the United Nations organisations, including WHO, ITU, UNESCO, UNDP and The World Bank, promote and collaborate in the development of TeleCentres serving a range of purposes, from basic telephony for the village population to telematics support for various sectors. Some of the first few such TeleCentres have been developed in the proximity of, or back-to-back to, the local clinic.

Throughout the above text, and in any and all published and unpublished writings that we have made on TeleHealth and TeleMedicine, we insisted on writing it as such – that is with a capital 'H' and a capital 'M'. This is to stress our firm belief that telematics in health will in the near future – certainly well within five years – be so extensive and so routine, that 'Tele' can be, and will be, readily dropped to emphasise the basic objective of Health and Medicine. How often do we hear these days the word 'microcomputer' used to refer to a desktop or laptop?

This book is primarily about eHealth in Europe and comprises superb contributions from European experts in and around eHealth. Many of the ideas and remarks I make in this Foreword are expertly elaborated in the 17 chapters of this book. Even though the contributions reflect European experience by European institutions, this book's contents could be true of, and would apply to, other regions of the world. The contents of the various chapters of the book also clearly affirm the global similarity of the findings to date of the practical experience in eHealth, particularly TeleHealth, and point to the importance of national and international collaboration for the resolution of issues of eHealth governance.

Salah H Mandil PhD
Director-Adviser
Health Informatics & Telematics
World Health Organization
Geneva
Switzerland
June 2000

List of contributors

Jonathan Bamford

Assistant Data Protection Registrar
Office of the Data Protection Registrar
Wycliffe House, Water Lane, Wilmslow, Cheshire SK9 5AF, United Kingdom

Jonathan Bamford has been with the Data Protection Registrar since the office was established in early 1985. He is currently Assistant Registrar, heading the Compliance Group with responsibility for ensuring that organisations in the police, criminal justice, local government, central government, education and health sectors comply with the requirements of the Data Protection Acts of 1984 and 1998. Prior to joining the Registrar's staff he was employed by the Equal Opportunities Commission advising on Sex Discrimination and Equal Pay cases.

Katherine Birch BA PhD

Research Fellow
Centre for Health Planning and Management
Darwin Building, Keele University, Keele, Staffordshire ST5 5BG, United Kingdom

After graduating in English and Sociology, Katherine Birch's early career was in NHS management, and she moved from this into public health research and service evaluation. Her PhD focused on the impact of central policy on the provision of maternity care as perceived by the users of such services, and she has always maintained a keen interest in the impact of service reconfiguration at an organisational, provider and patient level. She is currently a Research Fellow in the Centre for Health Planning and Management with responsibilities for postgraduate teaching and research, and works on a range of projects concerned with quality, the management of human resources and community and user involvement in the planning and delivery of healthcare.

Stephanie Bown LLB MBBS MRCP DRCOG

Medico-legal Adviser
Medical Protection Society
33 Cavendish Square, London W1M 2PS, United Kingdom

After qualifying in medicine from the Royal Free Hospital Medical School, Stephanie Bown initially worked in hospital medical posts for five years, and was then a principal in general practice in south-east London for three years. She then joined the Medicines Control Agency as Medical Assessor to the Committee on Safety of Medicines and the Medicines Commission, before joining the Medical Protection Society in 1994. She is a qualified mediator, and her particular professional interests include disciplinary procedures, advocacy and medical education.

Göran Carlsson MD

Senior Medical Adviser
Västernorrland County Council
Department of Health Policy, Västernorrland County Council, S-871 85 Härnösand,
Sweden

After graduation from the Karolinska Institute, Stockholm, Göran Carlsson spent a
working year in East Africa and the Middle East. Specialising in family medicine, he
subsequently held a position as a GP and later as District Medical Officer in Västernorrland, North Central Sweden. During the last 15 years, Göran Carlsson has been
working with health promotion and healthcare planning at the regional level; he has
also been the medical expert to InfoMedica, the Swedish non-profit producer and
distributor of a new patient information website.

Jari Forsström MD PhD

Director
Medical Informatics Research Centre in Turku (MIRCIT)
Kiinamyllynkatu 4–8, FIN-20520 Turku, Finland
Chief Physician
Atuline Virtual Hospital
Finland, www.atuline.com

After qualifying as a doctor and licensed physician in Turku University, Finland, Jari
Forsström obtained his PhD on the subject of 'Machine learning in clinical medicine by
knowledge acquisition from patient databases'. He was appointed Specialist in Internal
Medicine at Turku University in 1994, and two years later was appointed to his
MIRCIT post and a related position of Associate Professor (Docent) in Medical
Informatics at Turku University. He has been active in developing issues of health
informatics in Finland, including leading the European Union project Towards European Accreditation and Certification of Telematics Services for Health (TEAC-Health),
and the establishment of the Atuline Virtual Hospital. He has published over 30
international and 40 national publications related to medical informatics, legal and
ethical issues of telemedicine and quality of health-related Internet services.

José Garcia de Ancos LMS MSc (Econ)

Education and Information Adviser
British Medical Association
BMA House, Tavistock Square, London WC1H 9JP, United Kingdom

From a background in clinical general practice José Garcia developed an interest in
comparative health policy research after completing an MSc at the London School of
Economics and Political Science. Following various clinical academic appointments in
primary care in London he consolidated his previous experience in health service
management as medical adviser and director of service development for Ealing,
Hammersmith and Hounslow Health Authority; subsequently he has explored the
impact that different forms of delivery of clinical information have on the practice of
medicine as a result of the expansion of electronic media and the Internet. He is now a
senior policy adviser for the British Medical Association on those aspects of UK
government IM&T policy which have an impact on the access to clinical information
by doctors and patients in the NHS; he is a BMA representative on various working

groups to do with the implementation of the NHS Electronic Health Record and the Information for Health Strategy; he provides editorial assistance to *Clinical Evidence*, a BMJ compendium of evidence-based healthcare; and he is BMA secretary of the *British National Formulary (BNF)*, a joint publication of the BMA and the Royal Pharmaceutical Society of Great Britain.

Lorraine Gerrard BSc PhD
Research Fellow
Health Services Research Unit
University of Aberdeen, Foresterhill, Aberdeen AB25 2ZD, United Kingdom

Following a first degree and PhD in Pharmacology, Lorraine Gerrard remained at Dundee University transferring to the Department of General Practice. Here she investigated out-of-hours care within Tayside Region where telemedicine was used as one of the diagnostic options. Moving to the Health Services Research Unit in Aberdeen enabled her to continue research into telemedicine and the nursing profession. She is now a clinical trial co-ordinator researching osteoporosis.

Adrian M Grant MA DM BCh MSc MFPHM FRCOG FRCE
Professor and Director
Health Services Research Unit
University of Aberdeen, Foresterhill, Aberdeen AB25 2ZD, United Kingdom

After education at Oxford University, St Thomas' Hospital and the London School of Hygiene and Tropical Medicine, Adrian Grant's initial research career was at the National Perinatal Epidemiology Unit at Oxford; while there he co-ordinated a major programme of perinatal multicentre randomised controlled trials. In 1994 he moved to direct the Scottish Executive-funded Health Services Research Unit, a 50-person research group based in the University of Aberdeen. One of his main interests is the assessment of non-drug health technologies, such as telemedicine, including the impact they have on those who use them.

Judith Greenacre MB BCh DCH MFPHM
Consultant in Public Health Medicine
Dyfed Powys Health Authority
St David's Hospital, Carmarthen SA31 3HB, United Kingdom

Judith Greenacre is a graduate of the Welsh National School of Medicine who gained experience in general practice and child health before training in public health medicine. Now working in rural mid and west Wales, her key interests are child health and the development of practical information management systems. Over the past two years she has helped develop a successful congenital anomaly database for Wales (CARIS).

Ulf Marriott Hansson Leg Läkare
Physician
Department of Dermatology, Sundsvall-Härnösand County Hospital, 85186 Sundsvall, Sweden

After 10 years working in the pharmaceutical industry as a quality engineer, Ulf Marriott Hansson entered Medical School at Uppsala University as a mature student.

He completed his internship at Kiruna Hospital, and after a short period of locum work in general practice started his five year training in dermatology at Sundsvall-Härnösand County Hospital. During this time he has been involved in research into measurement of skin barrier function in hand eczema and has worked with the preparation for, and implementation of, a telemedicine project in the county of Västernorrland.

Irma Iversen RN JD

Patient Ombudsman
Akershus County
Pasientombudet for Akershus Fylkeskommune, Schweigaards Gate 4, 0185 Oslo, Norway

After graduating as a registered nurse in Denmark, Irma Iversen went to Norway where she graduated as JD at the University of Oslo, Faculty of Law. Afterwards she performed scientific work at the Norwegian Research Center for Computers and Law, University of Oslo, and later as an assistant professor at Aker Nursing School in Oslo. Since 1989, she has worked as Patient Ombudsman in Akershus County in Norway. She was the chairperson for Working Group V (Nursing) in the European Federation for Medical Informatics (EFMI), and a member of Working Group IV (Data Protection) in the International Medical Informatics Association (IMIA). Her main objective has always been to take care of patients' interests in the field of medical informatics.

J Ross Maclean MB ChB MD MBA MSc

Assistant Professor and Director of Health Services Research
Department of Medicine, Medical College of Georgia, Augusta, Georgia, USA
Vice President
Kerr L White Institute for Health Services Research, Decatur, Georgia, USA
Research Health Scientist
Augusta VA Medical Center, Augusta, Georgia, USA

After gaining a medical degree from the University of Aberdeen, Ross Maclean pursued his interests in remote healthcare working for the British Antarctic Survey and the Robert Gordon Institute of Technology Survival Centre, Aberdeen. This led to an interest in the role and evaluation of innovative health technologies in remote healthcare. After receiving formal training in Health Services Research (HSR) at the University of Aberdeen, Dr Maclean took a post to establish a program of HSR at the Medical College of Georgia. His research interests are health technology assessment, telemedicine, healthcare professional accountability and continuous quality improvement in healthcare.

Gustav Malmquist RN BSc

Director of Information Technology
Västernorrland County Council
Department of IT, Västernorrland County Council, Storgatan 1, SE-87185 Härnösand, Sweden

After 10 years practising as a nurse in ophthalmology at the hospitals of Sundsvall and Härnösand, Gustav Malmquist undertook research on the economic effects of co-operation between primary and secondary care levels. He is Project Manager for the telemedicine project in Västernorrland, and since 1998 he has been Director of Information Technology.

Frances Presley MPhil Dip Lib

Policy Officer
Association of Community Health Councils for England and Wales
30 Drayton Park, London N5 1PB

After graduating in American and English Studies, Frances Presley trained as a librarian, after which her first permanent post was as Information Officer with the Community Development Foundation. She moved through other information posts into the health field, including a period as Information Development Officer with the national SHARE project (Services for Health and Race Exchange) at the King's Fund. She is now Policy Officer at ACHCEW, the body which represents patients' interests in the NHS.

Åke Qvarnström

Telemedicine Project Manager
Västernorrland County Council
Department of IT, Västernorrland County Council, Storgatan 1, SE-87185 Härnösand, Sweden

After medical studies in the university town of Lund, Åke Qvarnström was appointed as a general practitioner at a primary healthcare station in Kramfors in the middle of Sweden. He has had an interest in telemedicine since 1996, and has been project manager for the Mid Sweden Telemedicine project in the county council of Västernorrland since 1998.

Michael Rigby BA FSS

Lecturer in Health Planning and Management
Centre for Health Planning and Management
Darwin Building, Keele University, Keele, Staffordshire ST5 5BG, United Kingdom

After graduating in geography and economics, Michael Rigby's initial career was within the National Health Service, commencing with community health policy research in Cheshire, and moving through information and planning positions to become Regional Service Planning Officer for Mersey Regional Health Authority. He then transferred to Keele, and the linked activities of research, applied development work and postgraduate teaching. He has always had an interest in practical innovation in the use of information, being active in UK and European initiatives, and he is Series Editor of the *Harnessing Health Information Series*, published by Radcliffe Medical Press.

Ruth Roberts RGN, SCM, MSc

Lecturer
School of Postgraduate Studies in Medical and Health Care
Maes-y-Gwernen Hall, Morriston Hospital, Swansea SA6 6NL, United Kingdom

Ruth Roberts undertook her general nurse training at the Queen Elizabeth Hospital, Birmingham (where she was awarded the Gold Medal for general training), then qualified in midwifery in Bristol and Exeter, after which she specialised in care of the elderly, culminating in her appointment as Nurse Consultant (Services for the Elderly) for Cambridge Health Authority. She then moved to Llanelli, Wales, as project leader developing a prototype Clinical Nursing Information System, from where she went on to manage the national Nursing Information Systems project for Wales. In her current position she undertakes a range of teaching and research activities, including organising an MSc course in Clinical Audit and Effectiveness, and is involved in a number of projects in Wales, England and Europe. Her publications include the recent *Information for Evidence-based Care* (Radcliffe Medical Press, 1999).

Rosemary Taylor LLB
Director of Primary Health Care & Information
Tees Health Authority
Poole House, Stokesley Road, Nunthorpe, Middlesbrough TS7 0NJ, United Kingdom

Following qualification as a solicitor, Rosemary Taylor worked in private practice for a number of years, before moving into public/private sector business development. This led on to a development position in Cleveland Family Health Services Authority, working with primary care general practice until the amalgamation of district health authorities and the FHSA in Teesside, when she undertook the position of Corporate Affairs Director and lately Director of Primary Care & Information. One of her main areas of work is to encourage the development and use of information across the health and social care natural community of Teesside; she is involved at a national level in networking, security and confidentiality issues and Teesside is a national demonstrator site for *Information for Health*.

Michael Thick MA MB BCh(Hons)Cantab FRCS (Eng) FRCS (Ed)
Consultant Transplant Surgeon
Freeman Hospital, Freeman Road, Newcastle-upon-Tyne, NE7 7DN, United Kingdom

Michael Thick is a practising clinician with a lifelong interest in applying information systems. His particular interests are rapid analysis and design using the Unified Medical Language System (UMLS), theoretical and practical applications of access control which represent patient consent, and the use of clinical decision support systems (CDSS) within NHS Direct. He is also a Health Action Zone Fellow, and Chief Medical Officer of Access Health UK Ltd.

Paul Wallace MB BBS MSc FRCGP
Professor of Primary Health Care
Department of Primary Care and Population Sciences, Royal Free and University College Medical School, Royal Free Campus, Rowland Hill Street, London NW3 2PF, United Kingdom

Having trained initially in medicine and subsequently specialised as a general practitioner, Paul Wallace went on to take up an epidemiology research training fellowship with the Medical Research Council and obtained a masters degree in epidemiology at the London School of Hygiene and Tropical Medicine. Following experience with a large randomised controlled trial to test the effectiveness of GPs' advice for patients with excessive alcohol consumption (undertaken through the MRC General Practice Research Framework), Paul Wallace continued to develop his interest in general practice research. He was appointed as a senior lecturer at St Mary's Medical School where he headed up the Helen Hamlyn Research Unit on Elderly Care, and was appointed to the David Chair of Primary Health Care at the Royal Free Hospital School of Medicine in 1993. His research interests in general practice are various; he is Chair of the European General Practice Research Workshop; he is currently director of the North Central Thames Primary Care Research Network (NoCTeN); and the principal investigator with a multicentre randomised controlled trial and economic evaluation of Virtual Outreach being conducted simultaneously in central London and rural mid-Wales.

Petra Wilson BA, DPhil

Lecturer in Law/Scientific Visitor to the European Commission
Law School, University of Nottingham, University Park, Nottingham NG7 2RD, United Kingdom

After completing a doctorate at the Centre for Socio-Legal Studies, Oxford, on the socio-legal dynamics of HIV and AIDS, Petra Wilson took up a full-time lecturing post at Nottingham University. Through teaching healthcare law at undergraduate and post-graduate level in both the Law School and the Medical School, her research interests developed in healthcare law generally and specifically in issues of confidentiality and security of health-related data. Her research in the area led to a secondment to the Health Informatics Unit in the Directorate General of the Information Society at the European Commission, where she works as expert adviser on info-ethics in health.

Richard Wootton PhD, DSc

Professor
Centre for Online Health
The University of Queensland, Brisbane, Queensland 4072, Australia

Richard Wootton moved to the Centre for Online Health as Director of Research in 1999, having previously been Director of the Institute of Telemedicine and Telehealth at the Royal Group of Hospitals in Belfast, where he also held a chair at Queens University. He has a long track record of innovation and evaluation in telemedicine. He is Editor of the *Journal of Telemedicine and Telecare*, published by the Royal Society of Medicine in London.

Jeremy Wyatt DM FRCP FACMI

Director
Knowledge Management Centre
School of Public Policy, University College London, 29 Tavistock Square, London WC1H 9EZ, United Kingdom

After qualifying as a doctor and working in hospital medicine, Jeremy Wyatt trained in medical informatics and wrote his thesis about a randomised trial of a computer tool to assist the management of patients with chest pain. He was an IBM Research Fellow for two years, then spent a year at Stanford and five years at the Imperial Cancer Research Fund before moving to UCL. He is interested in health technology assessment and evidence-based healthcare and is exploring how the Internet will change health services.

Acknowledgements

*The material in this book is based on the outcomes of
an expert working meeting
sponsored by The Nuffield Trust,
and organised by the
Centre for Health Planning and Management,
Keele University
and the
School of Postgraduate Studies in Medical and Health Care,
Swansea.*

Introduction

Computer technology has set off an information explosion that will change our lives beyond recognition. History has never seen a revolution on the scale of the one caused by the computers today. Douglas Robertson, in his book *The New Renaissance: computers and the next level of civilisation*,[1] offers an important historical perspective on the computer revolution by comparing it with three earlier landmarks of human invention: language, writing and printing. Each of these inventions changed the way in which we produce, store and distribute information and each one triggered an information explosion that transformed human civilisation. But the electronic computer has touched off the largest information explosion yet.

This is very much echoed in the recent government publication *Modernising Government*.[2] There is a commitment to ensure that public services are available 24 hours per day, seven days per week. For example, by the end of the year 2000 everyone will be able to phone NHS Direct at any time for healthcare advice. Furthermore, there is a new and ambitious target for all dealings with government to be available electronically by 2008. Echoing the 'New Renaissance', the government's policy statements emphasise the information age:

> 'We will use new technology to meet the needs of citizens and business and not trail behind technological developments. We will develop an IT strategy for government which will establish cross-government co-ordination machinery and frameworks on such issues as use of digital signatures and smart cards, websites and call centres.'

Progress will be measured against benchmarks, including targets for electronic services. Information technology is seen as revolutionising our lives including the way we work, the way we communicate and the way we learn. The information age offers huge scope for organising government activities in new, innovative and better ways and for making life easier for the public by providing public services in integrated, imaginative and convenient forms, like single gateways, the Internet and digital TV.

The Prime Minister anticipated much of this when he spoke at the *All Our Tomorrows* conference on the 50th anniversary of the NHS. He said, 'The challenge is for the NHS to harness the information revolution and use it to benefit patients'. He went on to say that cardiac patients are already having their heartbeats monitored by telephone and that the day is not far off when the Internet and interactive television will give us the convenience of home visits through technology.

In the recently published Nuffield Trust Series No. 9, *Realising the Fundamental Role of Information in Healthcare Delivery and Management: reducing the zone of confusion*,[3] Michael Rigby and his colleagues have highlighted many of the challenges this 'New Renaissance' will face in the NHS, particularly in the light of the history of IT in the health service. Whilst the importance of healthcare information is unquestioned, there is only a limited consensus on its function or development. Instead of being a well-managed resource, information systems are currently fragmented. Tensions abound, caused by a conflict of objectives, priorities and methods. In his monograph, Michael Rigby seeks to reduce the 'zone of confusion' by calling for a new beginning, based on objective principles for developing information policy and applications. The essential supremacy of patient-based records is emphasised, defining their core function as a patient-focused information system. A revitalised scientific underpinning of developments in information commensurate with its fundamental role in healthcare is proposed, focused largely on the functional aspects. A corporate philosophy of innovation and shared learning is considered essential. By comparing the scope and likely fates of the present and previous NHS IT strategies, a recently published National Audit Office Report[4] also illuminates some of the coming hazards.

Telemedicine

The conference on the future of telematics and the various papers on telemedicine which are presented in this book deal with some of the major problems which lie ahead. The way forward for telemedicine lies not in the realm of scientific discovery or technological progress – we already have the intellectual know-how and the developing technology to apply telemedicine to healthcare. Telemedicine applications have, for example, been introduced with varying degrees of success into a number of sites both in the United Kingdom and elsewhere. Furthermore, administrative systems to support telemedicine have been developed and financing for their continuation sought. We have seen widespread advancement in the use of telemedicine particularly in Sweden and Australia, and some particular areas within the NHS such as dermatology, mental illness and NHS Direct, and there are clearly further and very immediate advancements of telemedicine applications.

The future of telemedicine is not dependent therefore on technological issues although continued development to keep abreast of changing medical practice and new technology is essential. Rather, the way forward depends on an awareness of human values and needs. Mark Fergusson, speaking at a conference to celebrate the 25th anniversary of the Office of Chief Scientist in Scotland in November 1998, said that unlike the past major bottlenecks the limiting factor to progress is unlikely to be scientific discovery rather than implementation and social acceptance, as well as the provision of the administrative and financial systems. Spilker goes further and has pointed out that 'simplistic, futuristic futures and projections that ignore human values and needs will almost always be wrong'.[5]

This needs-driven agenda encompasses four issues:

1 people need to be placed at the centre of deliberations about the introduction and development of telemedicine
2 the whole question of ethics needs further consideration
3 attention needs to be given to the configuration of services if telemedicine is to be widely introduced, and alongside this are the relevant policy implications that need to be considered, e.g. commitments to equity, access, personnel management, and so on
4 for telemedicine to develop, consideration needs to be given to the whole question of management and leadership in the health service, as well as in related fields.

Telemedicine requires a substantial re-evaluation not only of patterns of delivery but also re-evaluation of the fundamental concepts upon which the NHS has been founded: local delivery, professional hierarchies and so on.

Globalisation of modern-day life: the localism of telemedicine

We already use a range of electronic appliances in our everyday life: why is healthcare not part of this? Major world bodies such as the European Union and the World Trade Organization already have telemedicine on their agenda. Yet all too often telemedicine is treated with suspicion and its successes are not developed.

What we have witnessed up to now can only be described as a piecemeal or fragmented approach to the development and introduction of new technologies in healthcare. This has led to a high degree of localisation and decentralisation with little thought being given to the regulatory framework within which telemedicine applications are developed, or to the wider cultural organisational and human implications of their introduction.

Historically, the NHS does not have a strong track record of shared knowledge and experience. As with many developments in the NHS, telemedicine's development has often depended upon individual visionaries or local providers working with suppliers to develop and implement applications which meet local service needs. Whilst such efforts have provided some notable success, too often little attention has been given to aspects of quality assurance such as data protection, ethics, standards, liability and accountability. This is gradually changing and it is essential that shared development and common knowledge be encouraged.

Given the sensitive and personal issues surrounding the delivery of healthcare, one of the major concerns for any future development is ensuring that the needs of the people using the new technologies, both professionals and patients, are placed at the centre of any decision concerning their development and introduction.

Which way?

As part of a project looking at the future of the health service, the Nuffield Trust supported some work on the development of scenarios. There were two types. The first – the *Find My Way* scenario – places particular emphasis on individual self-reliance, individual knowledge and local developments and therefore local guidelines. The second approach – *Trust My Guidance* – looks to limit the activities of the individual through strong regulation, the development of standards and involvement of strong national government. That is not to say that individual innovation and development are necessarily stifled, rather that there needs to be a strong national and international evidence-based framework within which telemedicine developments occur. One of the major issues is striking a balance between these two approaches. Standards and a clear regulatory framework are required to protect both the professional and the patient. However, innovation in response to local service needs should be encouraged.

Irrespective of which approach predominates, a whole-life perspective needs to be incorporated into the package. That is to say, carers and users need information that is relevant across a broad range of services – social services, health services and local government. This again raises the question of compatibility between systems, issues of data protection and access, training and so on. It also involves an acknowledgement of the shared responsibility for health which is placed on all public sector bodies – the NHS, education, social services and local government. As we have already said, central to any telemedicine development are questions concerning ethics. Clinical ethics in the United States, for example, is far more advanced and we need to consider in much greater detail ethics as applied to telemedicine and the NHS.

Policy and organisational issues

Healthcare policy developments emerging from telemedicine will enable a greater concentration of expertise and equipment in a small number of specialist centres dealing with complex cases. And this will be encouraged by the increasing sophistication of medicine in areas such as genetics, biotechnology, bioengineering, image-guided surgery, robotics and transplantation. Then there is the growing importance of technologies which allow self-diagnosis and self-treatment. There will be a greater concentration of common conditions in small centres linked telemetrically to specialist centres. These new technologies are likely to improve the potential for screening and treating serious conditions which will lead to the move from a sickness service to the construction of a genuine health service in which disease management and prevention are accorded higher priority. We will see a blurred distinction between primary and secondary care with more complex care taking place at home and the decentralisation of laboratory technology. We will see further reduced lengths of hospital stay as people are enabled to have greater amounts of diagnosis, treatment and monitoring activity take place outside the hospital.

Managerial and leadership challenge

In summary, the major policy implications of telematics require not merely a reorganisation of health services but a fundamental redesign as outlined by Professor Morton Warner in the first of the new Nuffield Trust Series, published in 1998.[6] The other policy implications include sectoral boundaries and the debate on the appropriate balance between primary, secondary and tertiary care. There will be resource implications. Many of the new technologies will substitute for existing interventions but may still result in additional cost for the service, for example in terms of screening or the ability to treat previously incurable conditions. But information technology and decision support will be paramount.

The 'New Renaissance' and the Prime Minister's commitment to the information age will pose significant managerial and leadership challenges, requiring managers to be able to manage both converging and diverging systems. We will have to move from a blame culture to a learning culture; from transactional to transformational leaders; from managers who have an interest and a focus in operational management to those happy with wider strategic issues; and move from competition to co-operation; from part-by-part to an holistic outlook; from a secretive to an open approach; a shift from managerial objectives to managerial values; to accept comfort with ambiguity in place of certainty; to encourage managers who are not risk-averse to being effective risk managers; and to move from a command-and-control system to one which recognises a network of influences.

So, in conclusion, the NHS has been presented with the chance to put behind it years of confusion and failure in its dealings with IT. The technology is there, and it works. The government is as keen to see the 'New Renaissance' spread throughout the NHS as it could be. So resources and a lack of political will, the two foes of successful IT implementation over the last 20 years, have been, it would seem, vanquished. Now we must grasp the human issues that are, contrary to many lay-people's perceptions, absolutely central to the development of IT. How do we make a culture change of the size required by *Information for Health*[7] happen? How do we meet the concerns of patients and clinicians for the security of information? How do we, as a service and a nation, cope with change in the balance of knowledge between professionals and the public that IT will bring?

This book is a contribution to this debate. It is a debate worth having as much as any throughout the history of the NHS. If the health of this nation is not to be left far behind that of comparable nations, we must find the answers.

John Wyn Owen
Secretary
Nuffield Trust
June 2000

References

1 Robertson D (1998) *The New Renaissance: computers and the next level of civilisation.* Oxford University Press, Oxford.
2 Cabinet Office (1999) *Modernising Government.* Cm 4310. The Stationery Office, London.
3 Rigby M (1999) *Realising the Fundamental Role of Information in Healthcare Delivery and Management: reducing the zone of confusion.* Nuffield Trust, London.
4 National Audit Office (1999) *The 1992 and 1998 Information Management and Technology Strategies of the NHS Executive.* National Audit Office, London.
5 Spilker B (1991) Planning for medicine discovery in the distant future. *Drugs News and Perspectives.* **4**(7): 389–93.
6 Warner M (1998) *Redesigning Health Services: reducing the zone of delusion.* Nuffield Trust, London.
7 NHS Executive (1998) *Information for Health.* NHS Executive, London.

CHAPTER 1

Telematics in healthcare: new paradigm, new issues

Ruth Roberts, Michael Rigby and Katherine Birch

Introduction

There is certain to be a significant increase in the use of telemedicine applications in healthcare delivery as we move into the twenty-first century, but there is a related major concern that these initiatives will be policy or technology led rather than evidence based, and that they will be based on old paradigm structures without consideration of the effects and challenges of the new paradigm that they create. For this reason, in December 1998 the Nuffield Trust sponsored an expert working meeting at the Royal Society of Medicine to consider these issues. This book is based on the contributions presented there, including the valuable plenary discussion.

This chapter will set the scene and introduce the topic of telemedicine. Chapters from the expert contributors at the working meeting then follow. The contributors of papers and participants in the plenary discussion raised several key points and issues which are incorporated into Chapter 16, and the concluding chapter looks towards the twenty-first century and the need for the potential significance of the effects of telemedicine to be considered in a global context as well as by individual nations.

The virtue in virtuality

As we enter the twenty-first century, it is self-evident that telecommunications will play an increasingly major part in the way that society functions and in how we live. The virtual organisation has already become part of daily life even before the ordinary citizen has become familiar with the concept. Telephone banking, telephone retail-selling, direct insurance, and centralised hotel bookings are regular daily transactions, yet the customer has no idea in what location – or indeed in what country – the operator is located.

Individual retail transactions are communicated by electronic point of sale (EPOS) systems not only to the retailer's headquarters, but onwards to warehouses, transport companies, product suppliers, and indeed to the production line – all part of a virtual organisation designed to ensure that a replacement product is available for tomorrow's customer. In turn, new industries have sprung up to the benefit of the consumer and society in general: these include telephone call centres which have brought employment to remote areas, and 'tele-cottaging' whereby staff can work from their own homes at hours which suit their other commitments. Transport companies have been transformed into logistics companies, changing their role from moving crates to collecting products from suppliers and distributing them to the stock-rooms of retail outlets on a top-up basis, refocusing their activity on the new function of managing the integrated information flow from point of sale to source of supply.

The underpinning communications concept, namely the telephone, is not itself particularly new. The major change has been due to rapid technological developments in terms of telecommunications capacity, data-processing and specialist peripheral equipment (e.g. automated tills), coupled with new organisational and transactional concepts. Consumers are the prime beneficiaries in terms of greater convenience and improved quality, timeliness and cost-effectiveness of services, and most new virtual retail and service organisations have grown directly as a result of consumer expectations, which in turn generate commercial rewards. More recently, images and electronic data interchange have started to replace voice communication in the form of Internet retailing.

Putting the tele- into healthcare

Healthcare is one of the largest sectors in any society, and it is of prime importance to both the citizen and the community. Within healthcare, communication – in the broadest sense of the word – is a vital component, and its timeliness, quality, availability and cost are all key factors in determining the quality of care. Therefore it would seem entirely appropriate that the healthcare sector should be at the forefront of harnessing the new telecommunications technologies, and that in turn healthcare delivery structures should be reviewed in order to maximise the opportunities of new communication modes.

The new health telematics vocabulary

This is indeed happening, and new terms are entering the vocabulary, matched by new areas of expertise. The specialist field of *health telematics* is defined by the World Health Organization (WHO) as a composite covering 'health-related activities, services, and systems carried out over a distance by means of information and communications technologies'.[1] Within this large domain *telemedicine* is the harnessing of health telematics for the delivery of healthcare services over a distance by individual clinicians. *Telecare* is the delivery of a broader range of care and support, including monitoring the daily living of individuals at risk, by harnessing telecommunications to other technologies ranging from television cameras to alarm systems.

But health is different

In principle, this process of ensuring that healthcare delivery continues to exploit the latest technological developments is surely to be welcomed. Indeed, it is simply the continuation of a long tradition of health professionals using new communication technology. For instance, the telephone has been an essential healthcare tool from the time when it first became available, several decades before it became a facility in the average domestic home.

However, health has its own very special characteristics, created by its essential importance to the individual and its intense personalisation, and compounded by its subjective and experiential nature. Consequently, the data used and the processes involved are frequently very personal, and of emotive relevance to the individual. This particular sensitivity (in all senses of the word) of personal health information means that the new communications technology needs to be harnessed very carefully.

The use of facsimile (fax) machines is a good example. Whilst the misdelivery of a motor-repair garage's order for vehicle parts would cause no more than minor annoyance to an unintended recipient or the original customer, misdelivery of a pathology report could have major and damaging consequences. Even the simple processes of arranging health referrals or reimbursement authorisation for clinical activity have necessitated the establishment of an organisational system known as 'safe havens', and call-back fax systems to ensure that delivery and confidentiality are protected. When patient contact is involved, although telephone consultation and advice can have many advantages, particularly outside consulting-room hours, an ineffective telephone contact (possibly with an unidentified individual) may be much less satisfactory and effective than a face-to-face visit.

Thus telematics in general, and telemedicine and telecare in particular, have tremendous potential to improve the quality and convenience of healthcare delivery, primarily by eliminating the barriers of distance. At the same time, there are major potential risks involved, ranging from loss of detail in the overall patient–health professional dialogue, through to overloading of key parts of the *ad-hoc* virtual health system, which could severely jeopardise quality.

The unique features of healthcare

This emergence of a new risk when healthcare harnesses telecommunications techniques is due to the special characteristics of healthcare functions, processes and organisations compared to other service sectors. In particular, healthcare is a highly customised person-to-person service, totally unlike banking or insurance, for instance, where the consumer selects products being offered or specifies the standardised services that they wish to be carried out.

In most commercial transactions the consumer is happy to deal with an organisation. They have no particular concern about where the telesales office is located, nor are they concerned that it may be an automated telecommunications system triggered by volume of traffic which determines the geographical destination of their call – they simply wish to benefit from a fast response. The identity of the person to whom they speak is of no major interest to them, a transaction reference being more important. Although the service or product that they are seeking needs to be personalised to them, this is on the basis of straightforward factual information which they are happy to supply, and for which they may well have a brochure or other guide to enable them to structure their responses.

Furthermore, the consumer benefits from the strength of organisational responsibility. The retail consumer has no interest in the complexity of the ordering and delivery processes triggered by the purchase of a similar item by another customer a day earlier, and involving a chain of interlinking independent organisations across several countries. Their only concern is the availability and quality of the product that they wish to procure. Therefore there is security in the legal and commercial liability of the retailer as the end-point of the virtual organisation, as it defines someone who must accept responsibility for and trace the origin of any problems in the timeliness or quality of service.

Healthcare delivery is entirely different in almost every respect. The very strengths of commercial systems become threats in healthcare.

Subtlety of health communication

The interpersonal communication between patient and clinician is subtle. It depends not only on dialogue, but also on general observation and specific clinical examination, whilst other senses, such as smell, may also be important. The dialogue will not follow a simple script, and the health professional is trained to look for non-verbal behavioural clues.

Personalisation of contact

In the delivery of care, the patient has little interest in the organisation, which indeed may be perceived as hostile. Rather, the interaction is extremely personal, with the identity of the health professional being seen as important, and continuity regarded as desirable. The consumer is reluctant to trust the impersonal healthcare organisation to deliver personalised yet confidential care.

The need for adaptation in application

Thus the effective harnessing of telecommunications into the new delivery mode of telemedicine and telematics-supported practice is important and potentially very beneficial, and is increasingly likely to occur. However, this cannot take place through a simple process of applying commercial techniques and methods to healthcare. Equally important, it cannot be achieved through the automation of existing modes of healthcare organisation and delivery.

Revised clinical skills are needed involving new communications and quality assurance processes, and the organisational effects also need careful consideration as patterns of referral may change dramatically. One important lesson which can be learned from the commercial world is that the escalating use of telemedicine, although it has the potential to bring major benefits, will necessitate new operational processes, and will result in significant restructuring of the healthcare organisation. In turn, this raises new issues of control and liability, requiring new referral and costing mechanisms and new educational techniques.

The importance of an evidence-based approach

Appropriately, healthcare policy and delivery are now placing great emphasis on evidence-based medicine (EBM) and evidence-based care. It would seem totally inappropriate to introduce an even more radical form of innovation without basing its adoption on firm evidence, and this evidence needs not only to be based on technical effectiveness in pilot sites, but also to include evidence related to the overall benefits and drawbacks in wider operational use.

The need for policy development

The World Health Organization has very valuably shown the way forward, and its own expert studies have identified the need for further studies and policy frameworks at global, national and local level as the prerequisite for safe and effective adoption of telemedicine applications.[1]. Roll-out or policy push of telematics and telemedicine without understanding, information, control mechanisms, and – above all – professional and public debate should be a major cause for concern, and Frances Presley raises serious questions concerning England and Wales in Chapter 14.

The need to recognise interests and stakeholders

Emphasis must also be placed on the need to look at health telematic applications in a multi-dimensional and holistic way. This has already been indicated in an international informatics setting with regard to telemedicine, as shown in Figure 1.1.[2]

There is a great risk of identifying benefits in one dimension whilst overlooking adverse effects in other dimensions. The ongoing consideration of telemedicine and other health telematics support as new means of delivering healthcare needs to include the holistic view of understanding the total health and healthcare situation. Figure 1.1 indicates that a range of patient interests (from both individual and societal views), health professional interests (separately from primary care and secondary care dimensions) and economic dimensions must all be considered and balanced. In addition, the overall effects on the healthcare delivery infrastructure and work-force should be considered, as part of a structured comprehensive assessment of all the potential benefits, offset by potential threats.[2] The importance of looking at the

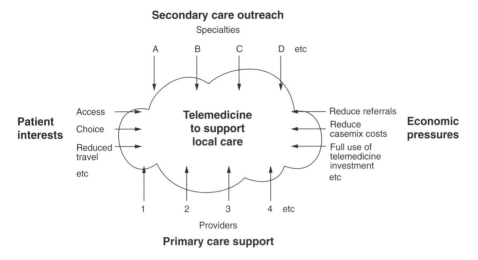

Figure 1.1: Stakeholder interests in telemedicine (after Roberts and Rigby).[2]

wider issues emerges repeatedly from the case studies and other chapters in this book.

The range of telemedicine applications

Although there are many important aspects of health telematics, telemedicine is the one which catches the imagination of the public and of politicians, as well as specialist scientists and clinicians. The range of telemedicine applications is large, growing and almost limitless. These applications can be grouped either by clinical domain or according to the different types of underlying telecommunications technology. An organisational categorisation could identify applications which automate the traditional referral processes up the hierarchy between primary and secondary care, and then between secondary and tertiary care (as illustrated by the Swedish case study described by Gustav Malmquist and colleagues in Chapter 3).

Continual creativity and challenged frontiers

Although these types of classification may be helpful to our understanding of the impact of telemedicine and its role in traditional healthcare delivery, they are by definition based on pre-existing models of service in which it is only the means of delivery which is new. However, new technologies only realise their true potential through radical thinking which abandons preconceptions whilst at the same time holding firmly to core values. Outside healthcare, for instance, major manufacturing clients have been able largely to abandon the traditional but expensive warehousing function as a result of informatics links direct from the production line to each component supplier's despatch base, enabling 'just-in-time' deliveries straight to the factory floor, whilst airlines can reduce costs and improve service quality by locating the sales-force in remote rural communities. Such radical rethinking is beginning to emerge in the healthcare sector with regard to telematics, and three examples are given below.

Healthcare direct

Recently, considerable emphasis and enthusiasm have been expressed by advocates of direct-access telephone advice services, such as NHS Direct in England[3] and similar services elsewhere.[4] In essence, these systems are designed to provide clinical advice for individuals concerned about a newly emergent health problem, and they are possible due to a combination of modern telephone functions and clinical decision support systems. Advocates regard direct services as a major improvement for the consumer through the provision of clinical advice when it is needed at any time of the day or night, while avoiding the problems of accessing general practitioner services (particularly outside normal consultation hours) and avoiding unnecessary hospital attendance and out-of-hours practitioner visits. Organisational proponents recognise the potential workload and cost savings at hospital emergency units and general practice surgeries, whilst ensuring that the public can obtain advice and, when necessary, immediate emergency care – in effect, a form of telematic triage. Critics see direct services as a further depersonalisation of healthcare, aimed at either reducing costs or avoiding otherwise necessary additional investment in traditional facilities. However, the effects on professional and public behaviour, and thus whether such changes are beneficial overall, have yet to be confirmed (although the NHS Executive is committed to rapid roll-out ahead of evaluative evidence).

Monitoring by telecare

The use of sensing devices can help impaired people in their daily lives – for instance, through such simple functions as switching off the light when a weight sensor detects that a person has climbed into bed.

Linked to telecommunications and a monitoring base, the same technology can enable vulnerable individuals to live in their own homes, as abnormal patterns of behaviour (e.g. leaving a cooker switched on) can trigger follow-up from a carer.[5] Such technology is already in place in Norway.[6] The individual surrenders a degree of privacy to the telematic surveillance system, in order to retain much greater privacy and autonomy than would be the case if they had to move to institutional or shared family living accommodation.

Electronic tagging

The technology developed to confine criminal offenders to their own homes could also be used to provide an element of safety within independence for people with early dementia or other confusional states, and could thereby improve their quality of life and avoid early institutionalisation by warning when the patient leaves their premises at untoward hours. However, the overcoming of stigma, not to mention a new ethical framework to consider the rights of adults who are not fully mentally competent, would be needed.[7]

Responsibility and liability

Issues of responsibility and legal liability need to be established in a telemedicine application setting. The clinical responsibility issue is regarded as more straightforward, as existing professional principles continue to prevail. As is shown in Chapter 3 from Sweden, and reiterated by the chapters that analyse the professional practice issues (particularly Chapter 9 by José Garcia de Ancos and Chapter 10 by Stephanie Bown), the test is whether the remote expert is directly treating the patient or whether they are advising the local clinician. If the actual treatment is delivered through the local clinician, the latter continues to have overall clinical responsibility. Whether or not they implement fully the advice received externally, and after what point – if any – they modify the recommended treatment in the light of other factors, is their personal responsibility. At the end of the day the traditional test of 'reasonable practice' continues to hold true. If the remote clinician takes over treatment directly, they must accept personal responsibility for ensuring that they have an adequate case history and relevant clinical details, and also that their treatment pattern has been administered directly or through a local agent according to their instructions. This area of legal issues in telemedicine is becoming an area of activity in its own right,[8] and is addressed and demystified in later chapters of this book.

Data protection

Although responsibility, liability and record-keeping principles may be clearer than anticipated, the issues of data protection are more difficult. Under the European data protection conventions[9,10] and resultant legislation in most European countries, a data subject must be aware whenever personal information is being recorded electronically, and must be in agreement with this. Moreover, other aspects then come into play, including the subject's access rights. Although it is self-evident that some form of electronic recording is highly likely to occur with teleconsultation and other forms of telemedicine, it is more difficult for the remote site to obtain formal agreement to this by the patient. In turn, the patient as data subject could experience organisational difficulties in exercising their subject access rights, particularly if the consultation is taking place in a different country.

Above all, however, there is a general failure by healthcare providers to appreciate that video clips and other forms of electronic image storage fall within the competence of the data protection acts as a form of personal electronic record. Indeed, where third parties appear in video clips (as is often the case), if they are identified as an individual (e.g. a named clinician or a named informal carer of the patient) they become data subjects in their own right, and are therefore themselves subject to the same legal rights and safeguards. Jonathan Bamford's contribution in Chapter 11 is therefore particularly significant, as in a domain which overall arguably has too little regulation, this aspect has precise legal requirements within Europe which are seldom fully appreciated.

Managing patient and clinician expectations

Little is yet known of patient expectations of telematics. In terms of knowledge, 'Internet syndrome' linked to consumerism is leading to increasing assumptions that the local clinician will be using the latest global evidence, without being able to define what that might be. For telemedicine, initial evidence (as, for example, in Chapters 3 and 7) indicates that patients welcome the elimination of travel and the enhanced speed of access, but this is in the atypical early-application situation with its implicit queue-jumping, and the enhanced clinical attention of two clinicians (remote and local) simultaneously. Whether this positive patient attitude will continue when telemedicine becomes part of the routine, with inevitable downgrading from its special status, remains to be seen.

In Chapter 7 Paul Wallace shows that clinicians are generally satisfied, although not to the same extent as patients – but that is to be expected of a more critically informed audience. Again the test will be to see how this holds up when telemedicine becomes routine in some settings, when hazards such as remote clinician non-availability begin to cause wasted time, and when e-mail and remote access systems have unplanned down-time or data loss.

Evaluation

The key to ensuring that telematics is harnessed and developed appropriately, and that a sound evidence base is developed upon which decisions can be based, lies in the development of better and longer-term evaluation methodologies. To date emphasis has been placed (rightly) on benefits assessment and benefits realisation,[11–14] but this is less than the full evaluation of all the direct and indirect results of a particular system implementation.[15–17]

Until now, the great majority of the evidence has been related to assessment of the technology and its immediate application – a critically important element, but only part of the story. The lack of wider evidence is a striking theme throughout the chapters of this book, and this echoes an analysis of the literature in 1995 which showed that only 15 out of 76 evaluative papers (20%) considered issues related to adoption and operational acceptance.[18] Telemedicine raises particular challenges, and therefore the contribution of Jeremy Wyatt in Chapter 5 is particularly important.

Recognising human and organisational factors

A strong emergent theme at the International Medical Informatics Association's last global Medinfo conference of the outgoing century was the need for the humanities to be given as much focus as the clinical and technical aspects in health informatics evaluation in general.[2, 19–21] There are two strong justifications for this. The first is that any branch of healthcare is about serving patients as people, and the second is that it is the personal and organisational behaviour of the deliverers of healthcare which determines how effectively or otherwise telemedicine or any other technology-based facility is utilised. In the meantime, however, the undue focus on technology alone has led to the understandable view that in health informatics research the human-ities are viewed 'like those big boxes with cheap socks and T-shirts just before an exit door in a department store' – a situation which cannot be either ethical nor effective in a people-based and people-focused service.[19]

There is empirical evidence from rural Scotland that, understandably, the adoption of telemedicine has been hampered by a perception that there is technology push, little integration into the overall health system, and lack of training.[22] However, there are now some moves being made towards looking more appropriately at the overall effects of telematics applications. Lorraine Gerrard and her co-authors in Chapter 8 highlight how staff who are expected to operate telematics systems feel excluded from the key decision making, and this is the latest example of an established phenomenon whereby the operational staff who are vital to successful information systems are the least valued and involved in the process of policy making and implementation,[23] thus creating a threat to overall success.

Integral and ongoing evaluation

Effective evaluation should not consider simply short-term studies, or technical evaluation of individual aspects of new developments, im-portant though these both may be. Reference has been made earlier to the importance of integrated and holistic evaluation whereby all dimensions are considered, and thus an overall assessment made.[2]

Secondly, evaluation should not be simply a 'snapshot' event. The case studies later in this book indicate how radical an effect telemedi-cine can have on the structure and pattern of healthcare delivery in a geographical locality, and these situations will become increasingly common and more widespread in the diagnostic areas and clinical specialities involved, as telemedicine approaches become increasingly utilised. Thus it is important for evaluation to be ongoing, and over time to progress from asking whether the system works to studying the effect on practice, and ultimately to assessing the effect on the healthcare organisation and its role,[24,25] as indicated in Figure 1.2. The need for a model-based approach to telemedicine evaluation has also been postulated.[26]

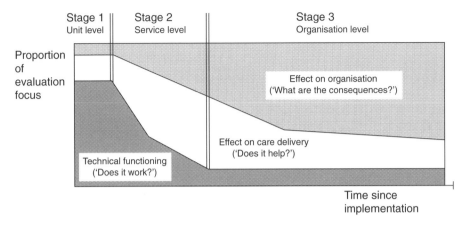

Figure 1.2: The necessary evolution of a system's evaluation (after Rigby).[25]

The globalisation effect

Technically, telemedicine knows no boundaries. A link can be as effective across a thousand miles as across one mile. Society is rapidly adopting a global approach, ranging from commercial out-sourcing to professional and private domestic use of the Internet.[27] At present, health telematics is often viewed over-simplistically as applying technical solutions, without realising the special characteristics of healthcare delivery structures. In the right settings and applications, bringing the 'tele' into healthcare clearly has advantages, and the examples and analyses described in the following chapters are intended to aid decision makers in reaching appropriate decisions.

Healthcare itself is currently rigorously controlled by national statute and regulation, and this will affect matters ranging from record-keeping to authorisation to practise, these requirements being a product of the perceived need to protect the citizen and to ensure professional competence. Thus, although telecommunications cross international boundaries effortlessly, delivery of healthcare in this way may not be lawful. For instance, in the USA many states require the registration of the health practitioner in the state where they are delivering care, so a national telemedicine service would require the clinician to be registered in every individual state. At the same time, existing European data protection legislation, although designed to protect the citizen, may not be framed appropriately to accommodate the citizen as patient.[28,29]

Conclusion

At present, health telematics is often approached over-simplistically as applying technical solutions without realising the unique character-istics of healthcare delivery structures. Second, applications are frequently introduced to improve healthcare delivery within existing healthcare provider structures (e.g. by removing travelling distance for patients, overcoming problems such as staff shortage, or enabling a planned working relationship with a more distant centre of expertise), but it is difficult to restrict these to operating only within the local setting, given global networking capabilities. Third, the Internet and other remote services are viewed as harnessing the most up-to-date global knowledge, but obtaining verification and authentication is difficult, as has been indicated.[30]

The answer to the question of how health telematics may be taken safely and effectively into the twenty-first century is a paradoxical one. The true opportunities will only be achieved by more radical thinking, as the industrial and commercial sectors have shown with regard to their own problems and solutions, but conversely the new problems and resulting challenges become easier to solve with reference to core and enduring health and professional practice principles.

Thus telematics in healthcare raises radical new opportunities and major new issues. The following chapters will explore both these dimensions.

References

1 World Health Organization (1998) *A Health Telematics Policy*. World Health Organization, Geneva.
2 Roberts R and Rigby M (1998) The need for a holistic view of telemedicine, focused on patients and society as prime stakeholders. In B Cesnik, AT McCray, and J-R Scherrer (eds) *Medinfo 98 Ninth World Congress in Medical Informatics, Proceedings*. IOS Press, Amsterdam, 1204–8.
3 Lister G and Flack I (1998) *Developing NHS Direct*. College of Health, London.
4 Lister G (1997) *Direct Health in Other Countries*. Coopers & Lybrand, London.
5 Fisk MJ (1997) Telemedicine, new technologies and care management (editorial). *Int J Geriatr Psychiatry*. **12**: 1057–9.
6 Kaasa K (1997) Implications of technology for the health and care services. In S Bjørneby and A van Berlo (eds) *Ethical Issues in Use of Technology for Dementia Care*. Akontes Publishing, Knegsel.
7 Fisk MJ (1997) *Electronic Tagging and New Technologies for Dementia Care: ethical dilemmas and good practice*. Paper presented at British Society of Gerontology Conference on 'New Thinking about Dementia'. Bristol, 19–21 September 1997.
8 Stanberry BA (1998) *Legal and Ethical Aspects of Telemedicine*. Royal Society of Medicine, London.

9 Council of Europe (1981) *Data Protection Convention*. Council of Europe, Strasbourg.

10 European Commission (1995) *Directive on the Protection of Individuals with Regard to the Processing of Personal Data and on the Free Movement of Such Data (95/46/EC)*. European Commission, Brussels.

11 NHS Information Technology Branch (1988) *Methods for Identifying the Costs and Benefits of Computer Systems Used in Health Care*. Department of Health, London.

12 Welsh Project Nurses Forum (1992) *Benefits Assessment Studies: a practical guide*. Welsh Health Common Services Authority, Cardiff.

13 Roberts R and Melvin B (1994) Benefits assessment. In SJ Grobe and ESP Pluyter-Wentig (eds) *Nursing Informatics: an international overview for nursing in a technological era*. Elsevier, Amsterdam.

14 Information Management Group (1998) *Benefits Management*. NHS Executive, Leeds.

15 Anderson JG, Aydin CE and Jay SJ (eds) (1994) *Evaluating Health Care Information Systems: methods and applications*. Sage, Thousand Oaks, CA.

16 van Gennip EMSJ and Talmon JL (1995) *Assessment and Evaluation of Information Technologies in Medicine*. IOS Press, Amsterdam.

17 Friedman C and Wyatt J (1997) *Evaluation Methods in Medical Informatics*. Springer-Verlag, New York.

18 Burghgraeve P and de Maeseneer J (1995) Improving methods for assessing information technology in primary care and an example from telemedicine. *J Telemed Telecare*. **1**: 157–64.

19 Hultengren E (1998) Health informatics and the humanities. In B Cesnik, AT McCray and J-R Scherrer (eds) *Medinfo 98 Ninth World Congress in Medical Informatics, Proceedings*. IOS Press, Amsterdam, 1180–3.

20 Lorenzi NM, Riley RT, Blyth AJC, Southon G and Dixon BJ (1998) People and organisational aspects of health informatics. In B Cesnik, AT McCray and J-R Scherrer (eds) *Medinfo 98 Ninth World Congress in Medical Informatics, Proceedings*. IOS Press, Amsterdam, 1197–200.

21 Hebert M (1998) Professional and organisational impact of using patient care information systems. In B Cesnik, AT McCray and J-R Scherrer (eds) *Medinfo 98 Ninth World Congress in Medical Informatics, Proceedings*. IOS Press, Amsterdam, 849–53.

22 Ibbotson T, Reid M and Grant A (1998) The diffusion of telemedicine: theory in practice. *J Telemed Telecare*. **4**: 1–3.

23 Finau SA (1994) National health information systems in the Pacific Islands: in search of a future. *Health Policy Planning*. **9**: 161–70.

24 Grémy F and Bonnin M (1995) Evaluation of automatic health information systems – what and how? In EMSJ van Gennip and JL Talmon (eds) *Assessment and Evaluation of Information Technologies in Medicine*. IOS Press, Amsterdam.

25 Rigby M (1999) Health informatics as a tool to improve quality in non-acute care – new opportunities and a matching need for a new evaluation paradigm. *Int J Med Informatics*. **56**: 141–50.

26 Bonder S and Zajtchuk R (1997) *Nuevo Paradigma para el Desarrollo y la Evaluación de la Telemedicina: Un Enfoque Prospectivo Basada en un Modelo*. Organización Panamericana de la Salud, Washington DC.

27 Rigby M (1999) The management and policy challenges of the globalisation effect of informatics and telemedicine. *Health Policy*. **46**: 97–103.

28 Rigby M, Hamilton R and Draper R (1998) Towards an ethical protocol in mental health informatics. In B Cesnik, AT McCray and J-R Scherrer (eds) *Medinfo 98 Ninth World Congress in Medical Informatics, Proceedings*. IOS Press, Amsterdam, 1223–7.

29 Rigby M, Draper R and Hamilton I (1998) The electronic patient record – confidentiality and protection of interests for vulnerable patients. In PW Moorman, J van der Lei and MA Musen (eds) *Proceedings of IMIA Working Group 17*, Rotterdam, 8–10 October 1998 (*EPRiMP: The International Working Conference on Electronic Patients Records in Medical Practice*, Department of Medical Records, Erasmus University, Rotterdam, 248–52).

30 Impicciatore P, Pandolfini C, Casella N and Bonati M (1997) Reliability of health information for the public on the World Wide Web: systematic survey of advice on managing fever in children at home. *BMJ.* **314**: 1875–9.

CHAPTER 2

The development of telemedicine

Richard Wootton

Introduction

Telemedicine means 'medicine at a distance'. In other words, it is a *technique*, not a technology. It is an umbrella term for many separate applications of medical care, including diagnosis and clinical management, treatment and medical education, whenever they are carried out at a distance. Similarly, telecare involves the provision of nursing and community support to a patient at a distance – for example, when the nursing staff are located at a different site to the patient.

The fundamental basis is the transmission of clinical information from one location to another, almost always by electronic means. For example, teleradiology involves capturing a digital X-ray image and transmitting it to a different site for display. Telepathology requires a system which can capture an image from a microscope, transmit it and display the image at a remote site. Teleconsulting (e.g. telepsychiatry) involves video-conferencing equipment installed at both the local site and the remote site so that the doctor and patient can see and talk to each other. The common thread is a client of some kind who is obtaining an opinion from a distant expert.

Telemedicine, or at least certain aspects of it, can be regarded as a subset of the entirety of medical computing ('telematics'), but it can also be regarded as a superset, as there is overlap between the two.

Telemedicine is glamorous. It is exciting to politicians and great things are promised for it, including the decentralisation of healthcare delivery beloved of policy makers, improved access for rural communities, accelerated referrals, better communication between primary and secondary care sectors, and reduced costs. However, as yet there is little evidence for most of these claims.

Background to telemedicine in the UK

Much of the early work in telemedicine was done in Scandinavia, where there has been a government commitment to ensure equal access to healthcare for the whole population, and where geographical barriers to travel often make this difficult to achieve.[1] In recent years there has also been an upsurge of interest in Australia and in particular in the USA, where some forms of telemedicine now represent major commercial activity (e.g. teleradiology) because of a similar need to make the best use of clinical expertise, in addition to major economic pressures.

The UK is a relatively late entrant to the field of telemedicine, and much of the work here is still at the experimental stage. Clearly the UK does not have the major barriers to access which exist in some parts of the world. However, the potential to improve healthcare in the UK by telemedicine is increasingly being recognised. To date this has mostly been the result of work by a small number of interested parties or telemedicine 'champions' who have been attempting to improve healthcare delivery using this technique. In 1998, however, the government announced plans to modernise the NHS and stated that information technology (IT) in general, and telemedicine in particular, will be important parts of this modernisation.[2] The government's intentions to introduce telemedicine into the NHS were clarified by the enquiry of Lord Swinfen, who asked in a Parliamentary Question for written answer:[3]

> 'Whether, if significant introduction of telemedicine (that is, remote diagnosis using information and readings supplied on-line or by telephone) is planned for the National Health Service, the fundamental principle of evidence-based medicine will be preserved; and whether:
> (a) telemedicine applications will only be introduced on the basis of identifiable clinical need supported by evidence of cost-effectiveness; and
> (b) external commercial pressures to introduce telemedicine will be resisted until evidence of cost-effectiveness has been obtained by scientific research trials in the National Health Service.'

The Parliamentary Under-Secretary of State, Department of Health (Lady Hayman), replied as follows:[3]

> 'The Government is committed to modernising the NHS, including introducing telemedicine applications where this is appropriate. These will only be widely introduced where there is clinical need and evidence from research and evaluation indicates that it is appropriate to do so.'

There is therefore a UK government pledge to introduce telemedicine as a means of delivering healthcare, but only where it has been shown to be an effective and efficient method of doing so.

Expected impact of telemedicine

In broad terms, telemedicine is simply a technique for medical communication which can be expected to improve the efficiency of a national health service by enhancing communication up and down the healthcare pyramid. Widespread adoption of telemedicine would permit significant decentralisation. Work that was previously carried out in the higher strata of the primary care sector could be carried out in the community, and work which had been the domain of the secondary care sector could be carried out by those in primary care. This would have an effect on hospitals, particularly at lower levels, although telemedicine can be expected to be an important survival mechanism for the specialist hospitals at the top, enabling them to export their specialist skills more effectively, as shown in Figure 2.1.

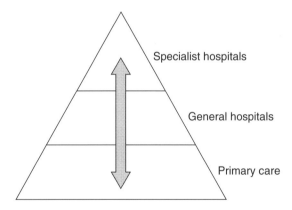

Figure 2.1: Telemedicine is a technique for communication in medicine. Improved communication up and down the healthcare pyramid permits decentralisation.

Telemedicine has obvious advantages in remote or rural areas where there are few specialist doctors. It can improve access to healthcare, reducing the need for patients or doctors to travel. However, even in urban areas the introduction of telemedicine can speed up the referral process, reduce unnecessary referrals, and improve the consistency and quality of healthcare.[4] Improved contact between the professional staff involved has been shown to produce both educational benefits and a reduction in professional isolation.[5]

What do we know about telemedicine?

Although it is growing rapidly, the literature on telemedicine is still relatively recent and one of its most striking features is how little formal research has been conducted so far. Whenever a new application for telemedicine is considered, it is almost always the case that little or nothing will be known about the underlying mechanisms (although, to be fair, any serious discussion about telemedicine will usually expose a comparable lack of information about the conventional alternative).

Telemedicine also seems to be unique in attracting instant 'experts' who profess strong, and usually mutually conflicting, opinions on its every aspect. (It is always worth applying the Yellowlees test,[6] and asking the current experts to define their practical experience of telemedicine, before listening to them too carefully.) This 'instant expertise' can be taken as an indication of the immaturity of the discipline. There is no general agreement either about the best applications or about the minimum technology required. In contrast, the pharmaceutical industry is about 150 years old, and it would be inconceivable that a new drug could be offered for sale without solid evidence of efficacy from a randomised control trial (RCT). Thus an RCT could be considered to be the bare minimum of proof required. Yet to date there have been virtually no RCTs in telemedicine, and as Hjelm has pointed out,[7] the basic evidence for cost-effectiveness is only now being gathered for a few selected telemedicine applications. This is important because few healthcare providers are likely to implement new techniques on a significant scale without solid evidence that they are at least as effective as traditional methods.

Although there are some studies of cost-effectiveness in the literature,[8] they are very few in number. However, irrespective of the lack of formal proof, telemedicine clearly works very well, both technically and clinically, given the right circumstances. In the UK, one of the more successful applications is decision support for nurse practitioners.[4,9] Teleradiology, a major commercial activity in the USA, will obviously become increasingly important in the future. In numerical terms the most significant current telemedicine application is the telephone helpline (note that in the UK this is planned to be introduced on a national scale in advance of any formal studies of its cost-effectiveness).[10]

Can we predict what types of telemedicine will be useful?

Given this background, can we predict what types of telemedicine will be useful in a national health service? The answer, I believe, is 'not accurately'. Another striking characteristic of the field of telemedicine is the rapidity with which new telemedicine programmes wax and wane. There have been very few such programmes which have operated for periods of longer than 5 years. The reasons for this are now beginning to be understood.[6] Although financial factors are undoubtedly important, it is the human factors which tend to determine the ultimate success or failure of a telemedicine project.

It seems likely that certain hospital-to-hospital applications will be useful, especially given the continuing government pressures to 'rationalise' the provision of hospital services. For example, there has recently been very positive experience reported in pilot trials of teleneurology.[11] Primary care telemedicine may be useful in certain applications (perhaps dermatology, for example[12]), although the economics have yet to be evaluated formally. Home telemedicine, which appears to meet consumer expectations of how the health services of the future ought to be delivered, may prove successful in due course.

Why research is needed

Research in telemedicine is urgently needed. There are two main reasons for this. First, the dearth of quantitative information about cost-effectiveness does not permit a rational debate about the merits and disbenefits of telemedicine. Secondly, it is not difficult to envisage a mistake occurring during a telemedicine episode which results in a legal action. Under English law and related jurisdictions, a doctor's defence can be that he or she had complied with a practice that was considered proper by a responsible body of medical opinion from individuals practising in that field (assuming reasonable skill and care were used).[13] Clearly, a responsible body of opinion would have to rely on published evidence of the efficacy of a new technique such as telemedicine – in other words it would have to rely on research data to indicate that it was safe and clinically accurate. Thus in the early stages of the adoption of telemedicine, critical evaluation of new applications is essential, as is the publication of the results of such evaluations.

For this reason, if for no other, it is important to publish the findings of experience with telemedicine, including case histories and other reports – even if these are negative – for the benefit of others, although the proliferation of purely anecdotal reports should be avoided. It is preferable to publish in the peer-reviewed literature.

Telemedicine research in Europe

In much the same way that the healthcare system in Europe is rather different to the healthcare system in the USA, telemedicine in Europe is at a rather different stage in its development. In the USA, telemedicine is a major growth area and some aspects have become significant commercial activities. In Europe, however, it is commonly believed that a more cautious approach is being adopted. None the less, or perhaps because of this, there is significant telemedicine research going on in Europe. In world terms, the two main regions from which telemedicine research is being reported are North America and Europe, as demonstrated by the analysis shown in Figure 2.2.

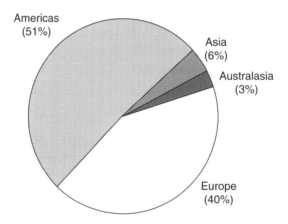

Figure 2.2: Primary telemedicine research. Origin of the primary research papers published in the two specialist telemedicine journals (*Telemedicine Journal* and *Journal of Telemedicine and Telecare*) 1995–1998.

What type of telemedicine research is being conducted in Europe? There have been some large and generously funded EC programmes, but these have mainly concentrated on 'high-tech' medical research, such as the use of advanced technology in medicine. Although considerable amounts of public money have been spent over the last decade or so, the work has been largely ineffectual in clinical terms. As has been observed before, the main problem in telemedicine is not a lack of new technology, but a lack of research on how to use the existing technology to the best possible effect. Very little work has been undertaken to evaluate the use of (existing) telemedicine technology in clinical practice.

In the case of individual countries, such as the UK, there is also a dearth of central funding for telemedicine work. The UK Department of Health has a health technology assessment programme, but so far it has taken little notice of the potential offered by telemedicine, possibly because this field is still too new and immature to merit serious consideration. Other countries, such as Italy,[14] have centrally funded telemedicine programmes.

In addition to formal research activities, much current telemedicine work can be regarded as informal research. For example, the US military are well known for their telemedicine activities,[15–17] but other European armed forces have also successfully experimented with telemedicine, and have published their findings.[18,19]

Funding for research

Governments naturally require scientific evidence of effectiveness as a prerequisite for the widespread introduction of telemedicine. Without robust, rigorously conducted trials, the widespread acceptance of telemedicine by healthcare professionals is also unlikely to be achieved. If telemedicine is to develop to its full potential and become generally more acceptable, it is essential that research is encouraged and that adequate funding is made available for that research. If telemedicine is to be developed effectively, then it is also essential that funding is made available for the areas of training and clinical service.

Research funding is therefore required at both national and local level, to encourage pilot telemedicine trials and definitive studies. For the most part, the areas to be studied should continue to be determined by those who actually deal with patients and their problems. Overall, the projects should aim to establish the benefits for patients, measure the reduction of costs and/or quantify the improvements to standards of healthcare.

Standards

Ultimately, as telemedicine matures, standards will be required for the technology itself and its operation. This includes protocols for the use of telemedicine techniques and training for the staff. However, it is a matter of debate whether such standards are required at present.

Over-emphasis on technical and data standards could be counter-productive in the long term, providing ammunition for sceptics, and possibly leading to over-ambitious and technologically driven initiatives. In the context of the latter, the NHS has a very poor record in implementing technology in general and information technology (IT) in particular. The portents for telemedicine are therefore almost uniformly bad. Standards for telemedicine should certainly be developed at the appropriate moment – but this may not be yet.

The issue of a national strategy

If telemedicine is worth introducing on a significant scale – which is an entirely open question in the light of current knowledge – then what is the right strategy? In many health services, the history of IT illustrates the dangers of starting with an expensive 'solution' and then looking for a problem to solve. In many industrialised countries telemedicine is largely being developed at the 'grass-roots' level by interested parties who in general are involved in projects relevant to their own specific area of interest. However, it has been suggested that a more co-ordinated approach might enable the potential of telemedicine for delivering healthcare to be realised more quickly. One way in which this might be facilitated would be by the development of a national strategy for telemedicine. Arguing against this there is the concern that the 'dead hand of central government' would stifle innovation and progress. Any strategy must therefore not consist of grand, immutable statements based on the misconceived notion that anyone knows where telemedicine will ultimately fit into the wider picture of health service delivery. Likewise, a strategy that places premature emphasis on such aspects as technical and data standards before the subject has even been allowed to mature – or even worse, the enforcement by centralised bodies of formal and restrictive evaluative approaches – will not only stifle the development of telemedicine, but possibly even marginalise the 'telemedicine champions' without whom it is difficult to envisage progress being made at all.

Instead, a national strategy for telemedicine must be flexible enough to reflect the ever-expanding pool of knowledge on the subject. At its heart it should promote telemedicine, support the evaluation of telemedicine applications through high-quality research, and eventually aid its implementation. However, for this to be achieved the strategy must not be a 'stand-alone' policy, but needs to be part of an established commitment by the government to:

- encourage and provide funding for telemedicine research

- develop a plan for implementation once effectiveness and cost-effectiveness have been demonstrated
- assess the major structural changes required within organisations to incorporate this method of delivering healthcare
- have a process for training, practice guidelines, quality control and continuing audit.

Telemedicine is clearly a technique that is worth investigating. However, it requires serious study, not a grand plan for implementation imposed from on high. The right approach to telemedicine is to 'start small', to prove the cost-effectiveness of a particular application in a pilot trial, and then to scale up cautiously to a wider service.[20]

Conclusion

Telemedicine has great potential for decentralising healthcare, for improving access and quality and (at least in principle) for reducing costs in certain situations. However, it also has a number of disadvantages. On the other hand, experience shows that telemedicine is certainly satisfactory in some if not many circumstances. The art lies in finding circumstances for which the advantages outweigh the disadvantages.

Given the right circumstances, healthcare services can be delivered cost-effectively through telemedicine, although there is little formal evidence available as yet, particularly in the UK. Having become technically and economically feasible, telemedicine deserves proper investigation. The development of telemedicine demonstrates a clear need for some proper research, and equally there is a clear need for proper research funding. Moreover, it is difficult to see how the UK government can achieve its policy (as quoted earlier) of introducing telematics applications based solely on research and evaluation evidence related to need if it does not facilitate the means of undertaking such research and collating the evidence.

References

1 Elford DR (1997) Telemedicine in northern Norway. *J Telemed Telecare.* **3**: 1–22.
2 Sherman J (1998) Blair promises eight years of higher NHS spending. *The Times* 3 July.
3 Hansard (1998) **593**: WA68–WA69.
4 Darkins A, Dearden CH, Rocke LG, Martin JB, Sibson L and Wootton R (1996) An evaluation of telemedical support for a minor treatment centre. *J Telemed Telecare.* **2**: 93–9.

5 Jennett PA, Hall WG, Morin JE and Watanabe M (1995) Evaluation of a distance consulting service based on interactive video and integrated computerized technology. *J Telemed Telecare*. **1**: 69–78.

6 Yellowlees P (1997) Successful development of telemedicine systems – seven core principles. *J Telemed Telecare*. **3**: 215–22.

7 Hjelm NM (1998) Telemedicine: academic and professional aspects. *Hong Kong Med J*. **4**: 289–92.

8 Allen A (1998) A review of cost-effectiveness research. *Telemed Today*. **6**: 10–15.

9 Tachakra S, Wiley C, Dawood M, Sivakumar A, Dutton D and Hayes J (1998) Evaluation of telemedical support to a free-standing minor accident and treatment service. *J Telemed Telecare*. **4**: 140–5.

10 Pencheon D (1998) NHS direct: evaluate, integrate, or bust. *BMJ*. **317**: 1026–7.

11 Craig JJ, McConville JP, Patterson VH and Wootton R (1999) Neurological examination is possible using telemedicine. *J Telemed Telecare*. **5(Suppl. 1)**: 137.

12 Loane MA, Corbett R, Bloomer SE *et al.* (1998) Diagnostic accuracy and clinical management by realtime teledermatology. Results from the Northern Ireland arms of the UK Multicentre Teledermatology Trial. *J Telemed Telecare*. **4**: 95–100.

13 Stanberry BA (1998) *The Legal and Ethical Aspects of Telemedicine*. RSM Press, London.

14 Pisanelli DM, Ricci FL and Maceratini R (1995) A survey of telemedicine in Italy. *J Telemed Telecare*. **1**: 125–30.

15 Gomez E, Poropatich R, Karinch MA and Zajtchuk JT (1996) Tertiary telemedicine support during global military humanitarian missions. *Telemed J*. **2**: 201–10.

16 Calcagni DE, Clyburn CA, Tomkins G *et al.* (1996) Operation Joint Endeavor in Bosnia: telemedicine systems and case reports. *Telemed J*. **2**: 211–24.

17 Norton SA, Floro C, Bice SD, Dever G, Mukaida L and Scott JC (1996) Telemedicine in Micronesia. *Telemed J*. **2**: 225–31.

18 Bilalovic N, Paties C and Mason A (1998) Benefits of using telemedicine and first results in Bosnia and Herzegovina. *J Telemed Telecare*. **4(Suppl. 1)**: 91–3.

19 Vassallo DJ (1999) Twelve months' experience with telemedicine for the British armed forces. *J Telemed Telecare*. **5(Suppl. 1)**: 117–18.

20 Wootton R (1998) Telemedicine in the National Health Service. *J R Soc Med*. **91**: 614–21.

CHAPTER 3

Health practice by remote expert: a case study from mid-Sweden

Gustav Malmquist, Ulf Hansson, Åke Qvarnström and Göran Carlsson

Introduction

What is telemedicine? How is it used in practice today? What are the benefits and what problems are connected with the implementation of telemedicine? These are questions that rightly need to be asked. The case of Mid-Sweden Telemedicine might provide some answers, but may also generate new conundrums that have to be solved.

What is telemedicine?

The simplest definition of telemedicine is 'distance medicine', which has been practised for almost as long as there have been ways to communicate over a distance. Patients are given advice from doctors and nurses every day by telephone. Physicians frequently exchange knowledge and expert advice by telephone, fax and also nowadays by email. In fact, it is possible to transmit medical knowledge by a huge variety of media, including Morse code and smoke signals, but it must be admitted that quality of care improves with the amount of information transferred. Pictures contain more information than a written description, but having both a picture and complementary information is even better. The ideal situation might consist of an image, medical data and the possibility of discussing this with the patient and the physician who has provided the information. This would amount to a fully interactive telemedical consultation. Undeniably the perfect situation occurs when doctor and patient can meet in real life, but under certain conditions this is not possible and telemedicine might then be the next best option.

The evolution of telemedicine

Even excluding simple technologies, the first trials with telemedicine were performed several decades ago.[1] However, for transferring images, medical data or interactive video, the technology used was insufficiently sophisticated and far too expensive. In more recent years this situation has changed as technology has improved and prices have fallen.

   The use of telemedicine as an aid to healthcare delivery or simply as a means of transferring medical information has followed the same pattern in Sweden as in the rest of the world. There have been trials using expensive technologies such as interactive satellite-television and there have been numerous pilot projects involving different types of image transfer over telephone lines, expensive data-transmission lines and the now more widely used integrated services digital network (ISDN). Most of these projects have focused on the technical aspect of transferring medical knowledge. Is it possible to make a correct diagnosis without being in the same place as the patient, but basing that diagnosis on electronically transferred images or examination data? As technology continues to improve the answer to this question is increasingly often 'yes'.

Teleradiology: the first full-scale application

Teleradiology was implemented at the four hospitals in Västernorrland County, Sweden, in 1994. According to a 1995 survey, teleradiology was in regular use at more than 70 hospitals in Sweden, and was thus the most widespread application of telemedicine at the time.[2] Consultations by video-conferencing were mainly used in scattered pilot projects. In a recent national survey it was confirmed that teleradiology is still the most frequently used telemedical application.[3] Interactive consultations using video-conferencing techniques are increasingly used, but mainly in pilot projects. However, video-conferencing is rapidly developing as a cost-effective alternative to distance meetings in healthcare organisations, and not only for medical use.

Mid-Sweden Telemedicine

Conditions for healthcare delivery

The County of Västernorrland is situated at the geographical mid-point of Sweden, with a population of 250 000 inhabitants in an area of 22 000 square kilometres. Community centres and industries are concentrated in the coastal region, whilst the inland areas are sparsely populated.

In Sweden, healthcare is chiefly provided by county councils. There is also a growing private healthcare sector that is most apparent in urban areas. Although the county councils provide most of the specialised care, local authorities are responsible for nursing homes and home-care delivery.

The County Council of Västernorrland runs four hospitals, namely Sundsvall County Hospital and district hospitals in Härnösand, Sollefteå and Örnsköldsvik. In addition, there are 40 primary healthcare centres (PHC) distributed across the county area. Most of the PHCs have between one and five GPs, together with health-visiting nurses, physiotherapy services and children's centres. A few PHCs provide visiting specialists as a service to their patients. Sundsvall County Hospital has most of the common medical specialities, and more specialised care is provided at the University Hospital of Norrland, in Umeå, 300 km north of Sundsvall.

Aims and goals

In 1995 a group of GPs and specialists started to discuss the idea of using telemedicine as an aid for primary healthcare in Västernorrland. Two years later, the Mid-Sweden Telemedicine project was launched with the aim of implementing telemedicine using video-conferencing techniques in a selected number of hospital departments and PHCs.[4] This would offer the following facilities:

- telemedical consultations
- second-opinion appraisal
- clinical and administrative conferences
- distance education
- project participation and work.

New ways of working as a result of the introduction of telemedicine were intended to contribute to a higher quality of care for patients who live in rural areas. It was hoped that they would also enhance the

Figure 3.1: Map of Västernorrland County in mid-Sweden.

availability of specialist advice and thereby lead to more rapid and flexible diagnosis and treatment and consequently to more efficient use of resources. Telemedicine might also increase the quality of care, as specialist services could be offered to those patients for whom travelling was difficult or impossible. In addition, it could serve as a tool to improve co-operation between primary and secondary care, which would obviously affect the quality of the process of care. The main expected outcomes of the project are as follows:

- increased quality of care in the selected districts
- lower expenses and less time spent on patient travel to hospital
- enhanced co-operation between different levels of healthcare providers
- improved competence of PHC physicians through the educational effect of consultations
- a higher proportion of relevant referrals
- a decrease in the rate of referral to specialists.

Project outlines

The project is divided into three sub-projects, namely telemedicine in primary care, telemedicine in occupational medicine and telepathology.

Table 3.1: Project phases and time schedule

Project phases		Activity
Phase 0	1995–96	Feasibility studies and pilot trial
Phase 1	1 January 1997–31 December 1997	Planning and purchase process
Phase 2	1 January 1998–30 June 1998	Main implementation period
Phase 3	1 July 1998–30 June 2000	Education, use and evaluation

Telemedicine in primary care

The project can be regarded as a pilot project at four PHCs, namely Ånge, Fränsta, Bredbyn and Kramfors, all of which are situated at a considerable distance from the nearest hospital. Ånge and Fränsta are situated in the southern part of the county, Kramfors in the middle and Bredbyn in the northern part.

Table 3.2: Participating PHCs

PHC	Distance to hospital (km)	Population served	Number of GPs
Ånge	100	6800	4
Fränsta	70	5000	4
Kramfors	40–90	9700	5
Bredbyn	40–200	5500	4

The main specialities that were focused upon in the primary care project were ear, nose and throat (ENT) and dermatology. The project is also prepared to involve other specialities such as psychiatry, physiology, orthopaedics and surgery. For specialist departments it was also expected that telemedicine consultations with other specialists could take place. Examples of such activities include second opinion and co-operation between specialist departments at different hospitals, as well as consultations with the university hospital in Umeå (300 km north of Sundsvall).

Table 3.3: Other sites

Specialist department	Number of physicians
Dermatology (Sundsvall)	7
Ear, nose and throat (Sundsvall)	11
Ear, nose and throat (Örnsköldsvik)	3
Occupational and environmental medicine (Sundsvall)	2
Emergency (Sundsvall)	N/A
Emergency (Sollefteå)	N/A
Clinical physiology (Sundsvall)	4
Pathology (Sundsvall)	5
Other installations	Usage
Örnsköldsvik hospital	All departments (except ENT)
Sollefteå hospital	All departments (except emergency department)
Härnösand hospital (to be installed in 1999)	All departments
County Council Headquarters	Administrative

N/A, non-applicable.

Telemedicine for occupational and environmental medicine

Occupational and environmental medicine in northern Sweden is located in Sundsvall, with a subsidiary unit in Östersund in the County of Jämtland, at the University Hospital in Umeå and at Boden Hospital in the County of Norrbotten. The co-operation between these departments will be facilitated and thereby increased by the use of telemedicine.

Telepathology

The first aim is to hold clinical conferences about patient cases between the pathologists in Sundsvall and surgeons at the hospitals in Sollefteå, Örnsköldsvik and Härnösand. Particularly in the fields of tumour diagnostics and cytology, such co-operation is essential for diagnosis and treatment outcome. The equipment is also used for second opinion and co-operation with pathologists, primarily at the University Hospitals in Umeå, Uppsala and Tromsø, Norway. Within this project an image analysis and storage system will be utilised that can also serve as a useful tool in the day-to-day analysis activities.

Project implementation

Fränsta Primary Health Care Centre – the first pilot

In May 1996 the first video-conferencing system in the county council was set up with one site in Fränsta Primary Health Care Centre, 70 km from Sundsvall, and the other in the emergency department at the County Hospital. The technical equipment was PictureTel 1000, and ISDN 128 kbps was used for communication. Valuable experience was obtained even though the system was not frequently used.

About 10 consultations were made during the trial period of 2 years, mainly concerning dermatology or surgical issues. The main reason for the low number of consultations seems to have been that most doctors at the County Hospital in Sundsvall were not involved with or adequately introduced to the telemedical consultation technique. The equipment in the emergency department was also considered to be a joint facility to be used by several departments. Most of the possible consultation cases did not concern emergency situations and it was inconvenient for specialists from other departments to walk to the emergency department for telemedicine work, due to their heavy workload. When certain specialists occasionally used the equipment it was also considered to be hampered by the low quality of the live pictures. However, transferred still images were of good quality, but it was not possible to take pictures with endoscopic equipment or to get close-up pictures. As a result, the clinical usage was too narrow to achieve a large number of consultations.

Change of approach

Experiences from the pilot project between Fränsta and Sundsvall showed, among other things, that it can be difficult to achieve frequent use of telemedicine if organisational and user issues are not taken into account. User involvement in the decision process, as well as user introduction and education, appeared to be essential in order to gain a wider degree of acceptance. The full project was started bearing this in mind, and groups were organised accordingly.

Purchase decision process

It was decided that video-conferencing equipment should be purchased for all sites, and also endoscopy equipment for the PHCs and ENT departments. The main requirements were as follows, in order of importance:

- user-friendliness
- technical reliability
- flexibility
- operating expenses
- cost of investment.

Providers of endoscopy demonstrated equipment to a project group. A purchase committee then collected opinions from the users. Two brands were regarded as equal with regard to most requirements except for the cost of investment. Most endoscopic equipment was purchased from Olympus Optical and some complementary equipment was obtained from Storz AB.

The evaluation of video-conferencing systems was performed by use of a protocol after demonstrations of the equipment. Marks were assigned on several criteria using a 4–point scale as shown in Table 3.4.

Table 3.4: Evaluation criteria

User-friendliness	*Technical quality*
User interface and monitors	Image quality, sharpness
Management (buttons, etc.)	Image quality, colours
Ease of use for making a call	Sound quality, clearness
Image transfer	Sound quality, duplex function
	Image storage capacity

Four brands were evaluated, including several models of each brand. A decision had to be made as to whether to choose PC-based equipment or a Roll-About system (PAL technology). PC-based systems are more flexible with regard to storage capacity, application sharing and communication within PC networks. The evaluation criteria that were found to be most important for the users were picture and sound quality as well as ease of use. Storage capacity was of only minor importance. The choice of brand was Sony Trinicom 5100P (a Roll-About system). For the pathology department a solution was chosen consisting of an integrated telepathology and video-conferencing system from Bild-analyssystem AB.

Systems specifications and communication

The overall goal has been to purchase as similar equipment as possible, in order to avoid compatibility problems and facilitate staff training. Generally the communication is carried out through ISDN 384 kbps.

Peripheral equipment such as document cameras and video recorders was purchased for most departments, and image storage devices and video-printers were purchased for some departments.

Cost of equipment

The total cost of equipment and project expenses has been estimated to be £540 000, of which £215 000 were funded by the European Regional Development Fund (ERDF). The price of equipment was for a full set with endoscopy and peripherals at approximately £40 800. For a basic set with only a video-conference and document camera, the price was £17 000. In addition, there was a cost of £570 each for connection to the ISDN network. The monthly fee for ISDN is £40 for each site and the fee per minute is between £0.08 and £0.17, depending on distance. The per-minute fee has been lowered twice since implementation as a result of competition on the telecommunications market.

Training

Introduction and technical training

In 1997 a number of introductory seminars open to all future users of telemedicine were held. In connection with the delivery of equipment, the suppliers have held user-training sessions in order to enable the users to manage the equipment. This training lasted for about 2 hours for video-conferencing and half a day for endoscopy. This is regarded as sufficient in view of the fact that the equipment is user-friendly. However, most GPs are not used to endoscopy. The ENT departments will therefore continue to teach the technique to all staff involved.

Telemedicine methodology training

However user-friendly the equipment may be, the practice of telemedicine is not simple. In numerous telemedicine projects world-wide the frequency of use has been lower than expected.[5] It is essential that doctors feel comfortable with the use and ethics of telemedicine, and that there are no unsolved legal problems connected to the practice. A course in telemedicine methodology has been planned, partly financed by national healthcare development funds. The idea is gradually to develop a training concept for telemedicine methodology that could be used nationwide, with the aim of enhancing physicians' confidence in telemedicine and thereby increasing the frequency of its use. Each

Table 3.5: Systems specifications at each site

Site	Sony 5100P	Document camera	Olympus endoscopy	Sony DKR storage device	Video printer	S-VHS	Other	Comments
Ånge PHC	x	x	x	x		x		
Fränsta PHC	x	x	x	x	x	x		Endoscopy prepared for gastroscopy
Kramfors PHC	x	x	x	x	x	x		
Bredbyn PHC	x	x	x	x		x		
Dermatology	x	x			x	x		
ENT Sundsvall	x	x	x		x	x		
ENT Örnsköldsvik	x	x	x		x	x		
Sollefteå Hospital	x	x			x	x		Dual monitors
Emergency, Sollefteå	x	x						
Örnsköldsvik Hospital	x	x				x		
Härnösand Hospital	(x)	(x)				(x)		
Occupational medicine	x	x				x		Dual monitors
Emergency, Sundsvall						x	x	To be replaced, PictureTel now
Clinical physiology							x	PictureTel 1000
Pathology department		x					x	Bildanalyssystem

(x), to be installed.

course will be conducted via a number of seminars in which ethical issues, patient methodology and performance as well as camera and studio techniques will be addressed. In the intervening periods, the participants will take part in practical exercises in groups.

A professional interviewer was hired in order to identify training requirements. It was important that the interviews were not conducted by anyone from the project management group, as users must feel free to 'tell the truth'. In addition to actual need for training, problems experienced with the use of telemedicine were also recorded for evaluation purposes.

Curriculum

On the basis of user needs, the outline for a course in telemedical methodology is as follows.

1 Basic technical training:
  • how to handle equipment
  • making connections, transfer of images, etc.
2 Methodology at meetings and presentations:
  • distance communication, problems and solutions
  • how to plan and perform a presentation by the video-conference
  • judicial and ethical problems.
3 Methodology for patient consultations:
  • non-verbal signals, how to 'read the patient' over a distance
  • pros and cons of telemedicine
  • ethical issues
  • exchange of experiences, best practice.
4 Telemedical technology:
  • imaging techniques
  • problem-solving
  • how to get the best from the equipment.

Practice of telemedicine

Consensus on how to practise telemedicine

In consensus documents, specialists and GPs have agreed on which patients should be offered telemedicine consultations and which patients are unsuitable for such consultations. For instance, the suspicion of a malignant tumour should be handled by ordinary procedures. During the project there has also been co-operation between specialists and PHCs about treatment procedures, particularly in the field of dermatology. In addition, the documents also regulate when the specialist services will be available. Initially services will be available once a week, but emergency cases could also be seen. The documents will be revised during the project in accordance with increased experience and evaluation of results.

Figure 3.2: Telemedicine practice at the primary healthcare centre of Bredbyn. Photo: Leif Vikberg, Örnsköldsviks Allehanda, Sweden.

Judicial guidelines

In November 1998 the Swedish National Board of Health and Welfare published a letter of advice to provide guidance on how existing rules and regulations should be applied to telemedicine.[6] The Board regards the current acts of law to be sufficient, and there would not be any need for a specific 'telemedicine act'. In order to facilitate the use of telemedicine, the following set of guidelines was issued.

Responsibility

Telemedicine is an alternative to the ordinary referral procedure where, for example, the GP has responsibility for the patient until he or she has been taken care of by the specialist. However, in a telemedicine consultation the specialist has only limited information and access to the patient, and therefore the overall responsibility remains with the GP. The specialist acts as a consultant and is responsible for advice based on the available information. In the event of misdiagnosis the question of liability has to be decided on an individual basis.

Medical records

A much debated issue concerns how and by whom the telemedical consultation should be documented. The official guideline is that, after finishing the consultation, the consultant should send a copy of a note on what advice was given. The physician who asked for the consultation (i.e. the physician who has the main responsibility) should do the main record-keeping. The purpose of a copy of the specialist's advice is to avoid misunderstandings.

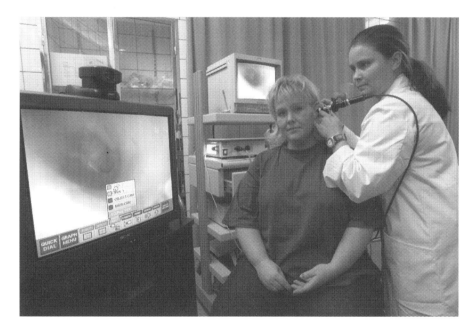

Figure 3.3: Two hundred kilometres from Bredbyn the remote expert at the Hospital of Sundsvall gives advice to the patient and her family physician. Photo: Leif Wiikberg, Örnsköldsviks Allehanda, Sweden.

Telemedicine in Sweden – a national perspective

Usage overview

In 1998 a national survey was conducted by the Swedish Health Planning Institute (SPRI).[3] Hospital directors and contact individuals from many telemedicine projects were asked about the ways in which telemedicine is used, and their beliefs on the potential and scope of telemedicine. Responses were received from 61 out of a total of 83 hospitals. Telemedicine is used in some form by almost 75% of all hospitals in Sweden. Teleradiology represents more than 20% of all applications described in the study and clearly it is now an accepted and widely used technology.

In the study, it was also shown that there are many projects in progress in the fields of ear, nose and throat, cardiology, neurology, ambulance care and dermatology. About 20% of the respondents considered that telemedicine can be applied to any speciality.

Definition of telemedicine from a practice point of view

According to the SPRI, telemedicine is a general term for transmission of medical information via telemedia. It is also recognised that it involves at least two people. It is therefore possible to state that the definition of telemedicine is concerned with the following:

- the technique used for transmission of information
- information contents
- who communicates with whom.

Telemedicine usage

Three main groups of usage have been identified, namely consultations, video-conferencing and education. However, it is obvious that whatever subdivision is used, there is an overlap between different types of usage.

1 Consultations (between two sites):
   - advice from specialist
   - sample results
   - on-call service
   - 'second opinion'
   - emergency/ambulance care.

2 Video-conference (several participants at two or more sites):
  - rounds
  - care planning
  - administrative planning
  - co-operation within a speciality.
3 Education (distributed from one site to another or several sites).

Participating levels of care

Telemedicine is most frequently used between hospitals (e.g. between a district hospital and a central county hospital, or between a central county hospital and a university hospital). As in Mid-Sweden Tele-medicine, the co-operation between primary care and hospitals is the point of interest in many Swedish projects. Telemedicine between nursing homes and primary care or a hospital is another level on which some telemedicine projects focus.

Telemedicine applications

It is significant for the currently rapid development of telemedicine that in the SPRI survey many projects had been started in 1998 or were about to be launched. The specialities involved in the projects are shown in Figure 3.4.

Risks and problems

In the national survey, judicial and ethical problems are highlighted as important in many answers. It is necessary to establish rules and practice as to who should be responsible for the patient in a telemedical consultation. It is also important to state how information should be recorded and stored after the consultation. Ethical risks that have been cited include, for instance, the fact that for more technical depend-encies there is a risk that the process of care might be impersonal. Many projects are facing obstacles in the form of rigid organisations, ignor-ance about telemedicine and a mentality of 'wait and see'.

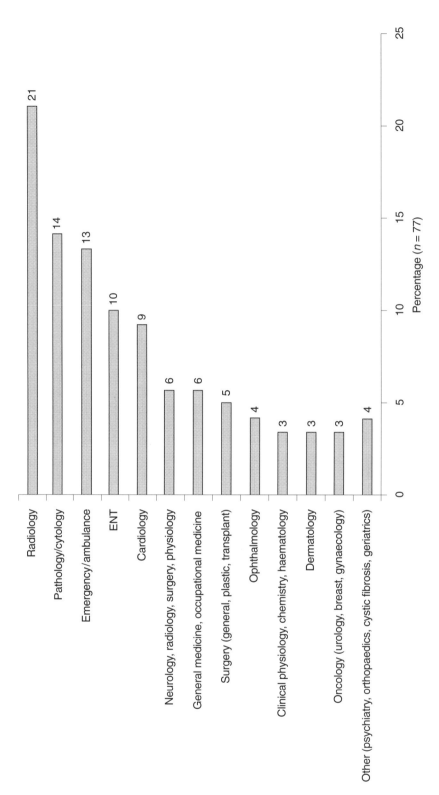

Figure 3.4: Telemedicine applications by specialty in the SPRI study in 1998 (after Holm-Sjögren *et al.*).[7]

Evaluation of Mid-Sweden Telemedicine

Evaluation is carried out in several sub-projects, and will focus on the following:

- technical and medical quality (dermatology only)
- economic effects
- organisational impact
- acceptance (attitudes).

It is important to have access to figures for comparison in a later stage of the evaluation. Before telemedicine was introduced in the departments, present and past performance was mapped with regard to number of patients, distribution of diagnoses (ICD-10), referral pattern and demographic data. This means that it would be possible to measure changes in, for example, referral pattern or how certain diagnoses are handled, as a result of telemedicine.

An easier form of evaluation that started once the equipment was in place involved recording use in a logbook. Cost of equipment is handled through the county council accounting system, and connection cost is measured by specified telephone bills for each site. Whether telemedicine is cost-effective or not depends not only on the cost of equipment or the cost of using it, but also on the frequency of use. If physicians and health professionals in PHCs and specialist departments accept telemedicine then it will be used, but if they do not the equipment will be yet another dust-collector.

Attitudes towards telemedicine

Before the installation of equipment the attitudes of all staff at eight of the departments involved were recorded (*n*=149). The survey also included questions about computer knowledge, previous experiences, and plans and ideas about uses for the equipment and how often it was to be used. There will be follow-up surveys including some of the same questions after 1 and 2 years of use. In this way it is hoped that changes in attitudes due to maturing of the technology and methodology in the organisation can be measured. Attitudes have been measured using a 5–step Likert scale. The question was: *'What do you think about the telemedicine that is about to be implemented in your department?'*

Table 3.6: Attitudes towards telemedicine by occupation

Percentage group	n	Great	Let's try	Don't know	Sceptic	Don't believe	Sum
Doctors	41	24.4	58.5	4.9	7.3	4.9	100
Nurses	38	2.6	55.3	31.6	7.9	2.6	100
Assistant nurses	16	18.8	50.0	25.0	–	6.2	100
Clerks	25	8.0	56.0	32.0	–	4.0	100
Others[a]	28	17.9	50.0	21.4	10.7	–	100
All[b]	148	14.2	54.7	21.6	6.1	3.4	100

[a]Others include, for example, midwives, biomedical analysts and physiotherapists.
[b]One missing occupation, response rate > 90%; n=149.

Results for first 9 months of use

During the period from March to November 1998 there were 129 telemedical consultations, 49 of which were dermatology cases and 60 of which were exercise ECG sessions by video-conference between Fränsta PHC and the Department of Physiology. Telemedical consultations for other specialities were in the fields of ENT (n=11), surgical and orthopaedic medicine (n=3), occupational medicine (n=3) and ophthalmology (n=1). In the field of dermatology, the number of consultations was very high compared to the number of ordinary referrals from the four PHCs per year, which in 1997 was 150. Three of the above consultations were to the University Hospital.

The use of telemedicine for educational purposes has exceeded expectations. A total of 17 educational sessions have been held from hospitals, PHCs, universities, university hospitals and other organisations. These do not include training in the use of telemedicine. Several events have been large-scale multi-point sessions reaching all sites in Västernorrland. Equipment has been used for meetings and research co-operation and supervision at least 40 times, especially in the fields of psychiatry and occupational medicine, where it is used on a regular basis.

Views and opinions gathered from interviews with telemedicine users[7]

Equity of care?

At the PHCs, telemedicine is regarded as a matter of equity for the patients. People who would have to travel a long distance for hospital care are given prompt access to the same service as is offered to urban patients. Telemedicine will also shorten the waiting-time from first contact with a GP to the final diagnosis. From the specialist perspective, telemedicine is regarded as more time-consuming than ordinary visits. Thus there is a risk that telemedicine might have a 'crowding out' effect on patients who are on the waiting-list from ordinary referrals. The question is raised as to who should decide which patients should be offered a telemedicine service and when.

Education and increase in competence

Telemedicine offers new opportunities for development and collegial contacts *for all categories of staff*. This is much appreciated at the PHCs. The same is true for specialists, although it is not specifically stressed.

Cost-saving or cost-driving?

GPs regard telemedicine as a way to achieve a correct diagnosis and treatment within a shorter time period, which would be cost-effective both for the patient and for society as a whole. However, some staff have questioned the cost and asked for less expensive alternatives to telemedicine, or for another use of the funds.

Rapid diagnosis or qualified guess?

GPs feel that they do indeed receive advice rapidly, and that treatment can be started earlier, but in some cases telemedicine also leads to referral of the patient. The likelihood of giving correct advice depends on both the GP's ability to take pictures of good quality and on other technical factors. Specialists consider that they make 'qualified guesses' more often than is the case when patients visit the specialist in person.

Change in working methods

Both specialists and GPs agree that telemedicine will lead to changes in routines and working methods. The GPs feel that specialists are somewhat hesitant about this possible development.

Judicial factors

During the interviews it became clear that some staff are uncertain about the rules that apply to telemedicine. This uncertainty concerns what is correct with regard to record-keeping, storage of images, responsibilities and confidentiality. However, others feel confident about the existing rules.

Discussion

Patient satisfaction

Mid-Sweden Telemedicine has not focused specifically on evaluation of patient satisfaction. However, a number of spontaneous comments made by patients have been generally positive about the telemedicine services. Patients appreciate that travel to hospital can be avoided, and in particular, that the consultation is conducted without days or weeks of delay on a referral waiting-list. They also express an increased sense of confidence and trust after having had the opportunity to discuss the problems with their GP and a specialist at the same time. The general opinion that patients are satisfied with telemedicine has been confirmed by numerous studies.[8,9]

Cost-effectiveness and technical solutions

Even if the project has been successful with regard to frequency of use in certain specialities and for certain applications, the number of consultations has to increase considerably before it is possible to say that it is cost-effective in this form. However, although expensive, the equipment chosen (Sony Trinicom) might have been a strategic choice, as the users feel that it is easy to learn to use, and they do not have to risk common computer failures. With the large screen and superior sound quality the contact between specialist and patient is also better than with a PC solution. However, the cost of equipment for a PAL system is much higher. In the long term PC systems may catch up in quality and offer more flexibility with regard to data storage, integration with medical-record systems and use of low-cost data-transmission lines. The cost-effectiveness will depend on many factors, such as which patients are taken care of by the use of telemedicine (travel costs), the value of other uses of equipment (e.g. education) and the degree of increased demand for services as a result of telemedicine.

Organisation

The results obtained from the first 9 months show that the rate of ENT use is lower than expected. Some training of family physicians in endoscopy technique has still to be carried out, which might be one explanation for this finding. The fact that telemedicine involves more than one doctor when it is used as a tool for specialist consultation will also lead to organisational challenges. If specialists are difficult to reach, or if there is a concern about lack of time at the PHC or specialist department, an ordinary referral might be too easy an option. In this project, telemedicine services are offered from the specialist department mainly once a week. Even if it is possible to make exceptions in certain cases, this means that the specialist department is not prepared for consultations at any time other than Wednesday morning, which is the time slot that has been chosen. If a GP contacts the specialist on some other day or at some other time in order to help the patient at the PHC immediately, this might be regarded as an intrusion at the specialist department. The reason for choosing a certain 'telemedicine day' was that the specialist departments have very long waiting-lists, fully booked out-patient clinics and sometimes staff vacancies as well, all of which result in a heavy workload.

However, a refusal of a telemedicine consultation will lead to increased use of healthcare resources, as the alternatives are for the patient to have either another appointment at the PHC (on a Wednesday morning) for a telemedicine consultation, or a referral to the specialist department without previous consultation. In addition to the inconvenience for the patient, with more travel involved, this leads to more planning and scheduling activities both at the PHC and at the specialist department. As specialists become more accustomed to telemedicine, it might be possible to handle telemedicine consultations like other unscheduled out-patients. The service could be handled by a fully qualified physician who is not fully booked up with appointments. This change of approach has been observed in a telemedicine project in the neighbouring county of Västerbotten.[10]

Attitudes

It is always difficult to change patterns of behaviour and attempts to try to change working methods in the healthcare sector are no exception. It is obvious that some doctors are enthusiasts, and for them there are often no problems. Most staff are indifferent to the concept, and the challenge is to engage those individuals and persuade them to see telemedicine as a natural alternative to the ordinary referral procedure.

According to Linderoth,[11] people have 'mental maps' of how health services should be provided. The structure of the organisation is built accordingly and is difficult to change if those maps are not altered. To put it bluntly, healthcare professionals have been trained to do things in a certain way, and they feel that this is the best way of doing them, since everyone else does them in the same way, too. If one tries to change this pattern it is not surprising that it is questioned or even criticised. The 'mental map' is transformed and staff have no readiness for alternative actions outside the familiar map. The need for time and effort to be invested in this organisational and mental change is probably very much underestimated. Telemedicine has often been regarded as a matter of technology, and it is thought that once the technology is developed the use of telemedicine will start automatically. It is forgotten that there are many people involved in the process of care, all with their own ways of thinking, their own tacit knowledge and their confident professional self-reliance. These people constitute the soul of the organisation, and without them, telemedicine will not be used frequently. To put it simply, telemedicine will not be cost-effective without an organisational change.

In the national survey by the SPRI, rigid organisations are the most frequently cited obstacles to success and reasons for difficulties in working with telemedicine projects.[3] It is possible that this rigidity includes both attitudes and formal organisational conservatism. Another problem is that there is a lack of time for doctors and healthcare professionals to participate in development projects, or to make the effort to try to change ways of working in the clinic. Hospital departments and PHCs in Sweden suffer from either strained staff situations as a result of limited budgets, or recruitment problems due to lack of trained physicians and healthcare professionals. The latter is particularly a problem in northern Sweden. In this situation, development projects such as telemedicine projects are not easy to manage, however important they may be if the hypothesis is that in the long term telemedicine will lead to more efficient use of healthcare resources.

There may be several ways to overcome obstacles to success. In the Mid-Sweden Telemedicine project the following issues have been addressed:

- user involvement throughout the process
- network-building between organisations
- consensus on usage and diagnostic criteria
- complete user-training, not merely how to press the button.

In most departments and PHCs included in the project users have been very active, but in some cases unexpected problems have arisen which are mirrored in the frequency of use during the first 9 months. Even if there is a problem of lack of time, it is essential that the actual users (i.e. not only physicians but also nurses and other healthcare professionals) are involved. When considering the introduction of telemedicine in a particular department it is important to ask not only 'Can we afford the investment?' but also 'Can we spend the time that will be needed for the organisational development?'.

Summary

The Mid-Sweden Telemedicine project started in 1997 and to date has implemented 15 telemedicine sites in the County of Västernorrland. During the first 9 months, 129 consultations were made. The frequency of use has exceeded expectations in the fields of dermatology, physiology and education. An innovative approach has also been adopted in the project, with new applications introduced, such as exercise ECG by use of video-conferencing techniques. One important conclusion drawn from the project is that telemedicine is more an organisational than a technical issue. If users are involved throughout the process, the potential for wider acceptance will be greater. Thus telemedicine will be able to be used both for the benefit of the patient and to enable better use of healthcare resources.

References

1 Bashshur RL (1995) Telemedicine effects: cost, quality, and access. *J Med Systems.* **19**(2): 81–91.
2 Olsson S (1996) *Telemedicin i Sverige, Projektkatalog.* SPRI, Stockholm.
3 Holm-Sjögren C, Sandberg C and Schwieler Å (1998) *Telemedicin: Kartläggning av tillämpningar i Sverige 1998.* SPRI, Stockholm.
4 Malmquist G and Carlsson C (1998) *Mid-Sweden Telemedicine, Project Report 1.* Report No. 35. The County Council of Västernorrland, Health Policy Department, Härnösand.
5 US Congress (1995) *Bringing Health Care Online: the role of information technologies.* Office of Technology Assessment, Washington DC.
6 Socialstyrelsen (1998) *Meddelandeblad Nr 12/1998.* Socialstyrelsen, Stockholm.
7 Gustavsson G (1998) *Notes from Interviews with Users of Telemedicine in Västernorrland.* Landstinget Västernorrland and TL Information AB (unpublished paper). Härnösand.
8 Bloom D (1996) The acceptability of telemedicine among healthcare providers and rural patients. *Telemed Today.* **4**(3): 5–6.
9 Linderoth H (1998) *Telemedicine: will noble ideas be translated into action?*

Umeå School of Business and Economics, Department of Business Administration, Umeå University, Umeå. Paper presented at International Telecommunications Society Twelfth Biennial Conference. 21–24 June 1998, Stockholm.

10   Jirlén L and Made C (1997) *Telemedicin i glesbygd*. Västerbottens läns landsting, Umeå.

11   Linderoth H (1997) *Telemedicin: Ädla idéer möter en bister verklighet*. Handel-shögskolan, Umeå University, Umeå.

Further reading

Liljenäs I and Strömgren M (1998) *Telemedicine in Marginal Rural Areas in Sweden: new opportunities for the citizens in the new millennium?* Umeå University, Department of Social and Economic Geography. Paper presented at PIMA98, 26–29 June 1998, Aberdeen,

Linderoth H (1996) *Adoption and Acceptance of Information Technology in Health Care Organizations: some theoretical and empirical findings*. Umeå School of Business and Economics, Department of Business Administration, Umeå University. Paper presented at INFORMS National Meeting, 3–6 November 1996, Atlanta, GA.

Malmquist G (1996) *Långt till sjukhus: ekonomiska effekter av decentraliserad specialistsjukvård*. Mid-Sweden University, Sundsvall.

CHAPTER 4

Remote electronic knowledge bases and information services: their compilation and quality assurance

Jari Forsström

Introduction

Information and communication technology (ICT) is rapidly changing the way in which medical knowledge is delivered and processed. In recent years, many pilot applications in telemedicine have shown the potential of telecommunication in medical consultations. However, electronic commerce of medical knowledge does not only involve consultation with human experts. It is possible to build computer programs, decision-support systems and knowledge bases that can be either purchased or consulted through the Internet. The great advantage of these applications is that they only require computer resources and bandwidth. As both telecommunication and computer costs will drop close to zero, the only costs for customers are those of the software or service in the computer. Since computer software is easily scalable, the costs of using the service will become cheaper as the number of users increases. This will lead to applications in very narrow fields of medical expertise. This new field of telematic health services is still in its infancy. Almost certainly the main obstacle will be the difficulty for the end-user in assessing the quality of the services that they find and thus the hidden risk they may pose. It seems evident that a third party is needed to give an objective judgement on the quality and accuracy of medical information and analysis. In this paper, a commercial application in clinical chemistry is described as an example. In the latter part of the paper, different ways of setting up third-party quality assurance are discussed.

Description of the example application

Background to the medical problem

The Drug Laboratory Interference (DLI) knowledge base is a remote knowledge base that is available on the Internet. The application is described to show how the Internet is increasingly being used in international collaboration to process and collect medical knowledge.

Many drugs may affect laboratory test results.[1-3] These effects often give rise to misinterpretation of clinical laboratory data and may lead to wrong diagnoses, and unnecessary treatments and laboratory tests. The number of registered human drugs is increasing rapidly. New laboratory tests are constantly coming into clinical use and new laboratory methods are being developed for old analyses. Therefore it has become very difficult to keep up with even the clinically most important drug effects on laboratory tests. As over 5000 human drugs and at least 500 laboratory tests are in routine use, together they represent 2.5 million possible combinations. Accordingly, it is not surprising that the effects of drugs on laboratory tests are not systematically evaluated. Most often the information is published in sporadic case reports involving a limited number of patients.

The medical literature has reported approximately 30 000 different drug effects on laboratory tests. However, the reliability of these publications varies widely. Some are based on animal studies alone, while some report the effects on a patient group but fail to document the effects on a normal population. Furthermore, there is often insufficient information about the nature of the effect. Some clinicians may have reported the effect as biological without taking into account the possibility that the effect may be analytical and specific to a certain laboratory test method. Accordingly, there may be contradictory published reports about a certain drug effect on a certain laboratory test, and thus an individual publication may be misleading. It is therefore important to collect all of the relevant and available published information on a certain effect in order to draw accurate conclusions with regard to current knowledge.

Several investigators have collected reference listings concerning drug effects on laboratory tests.[1,2,4,5] However, these data are unsuitable for online notification, as they do not include meta-analysis of the effects. International initiatives have been taken to harmonise the way in which drug effects on laboratory tests are evaluated and reported.[6] Although this work has made us more aware of the importance of the problem, it has had little effect on clinical practice. Usually only clinical laboratories are familiar with the drug effects on laboratory tests, and transferring this information to the wards has proved extremely difficult.

Drug Laboratory Interference (DLI) knowledge base

In 1991, the Medical Informatics Research Centre in Turku (MIRCIT) and the Central Laboratory, both part of Turku University Central Hospital, started a joint project known as CANDELA (Computer-Assisted Notification of Drug Effects on LAboratory test results). The aim was to provide automatic warnings to clinicians when a clinically significant drug effect on a laboratory test was detected.

It was evident that maintaining a knowledge base of drug effects on laboratory tests requires a strictly systematic way of describing the knowledge. As free text, rules are difficult to update and use in an automatic system. The essential features have to be coded in a manner that is suitable for computerised processing but at the same time, can be understood by a human expert. For this purpose, the Drug Laboratory Interference (DLI) code was developed. The first version was published in 1995.[7] The DLI knowledge base has been made available through the Internet, and the use of the Internet in collaboration between international experts is increasing.

The Internet as a tool for collaboration

In the information society, information and knowledge can be collected and processed effectively. The Internet has made global collaboration possible, enabling experts throughout the world to participate in joint research and development activities. Distance has thus lost much of its significance. Collaborating on the Internet requires new methods of working. As most participants do not know each other personally, there is no guarantee that the information collected is of high quality. Therefore the project needs one professional co-ordinator who is able to select the pieces of information to be included in the product. A good example of this approach is the Linux operating system developed by Linus Torvalds (an ex-student of the Department of Computer Science at the University of Helsinki). In 1990, he started to develop a Unix operating system for PCs, known as Linux. When the first version was ready, he decided to put the code on the Internet for free distribution. The code was found to be useful among computer scientists. Torvalds obtained assistance from the cyberspace community, and has selected the pieces of software for the final product.

In creating the DLI knowledge base, the development process of Linux was used as an example. The knowledge base was developed using a network of experts whose input was checked and finally approved for inclusion in the knowledge base by experts at Turku University. A PC version of the knowledge base is available free of charge at www.multimedica.com/dli

The DLI example is only one among hundreds of serious medical information services that have been made available on the Internet. Many of them are large collections of data, and it is impossible for the medical information to be checked by one person. As medical information services are not products in the classical sense of the word, it is difficult for the end-users to rely on these services. Building trust in these services is a major challenge for the future.

Evaluating the quality of telematic health services

Background

During the past few years, healthcare professionals' attitudes towards computers have become more positive. More medical doctors are involved in the development and planning of medical applications, and it is also recognised that the technology itself is no longer a problem. It is appropriate to use computers in healthcare to provide high-quality care at reasonable costs.[8-11] In addition, the initiatives with regard to medical software quality assurance have become more relevant from a clinical viewpoint.[12,13]

A year ago a small error in the software used for biochemical screening for Down's syndrome was detected.[14] It was a clear error, but fortunately it did not have a very dramatic influence on the probability estimations for Down's syndrome. Nevertheless, the error was reported in the *Lancet* and received wide publicity, clearly showing how important it is to evaluate thoroughly the medical knowledge included in software. Using computerised systems it is possible to make decisions about the treatments of thousands of patients a day. An error in a program can produce much more damage than the misunderstanding of a single physician, as the number of patients seen by one physician will always be very limited.

Lack of health telematics validation

Telematic applications have enormous potential for improving the quality of healthcare. However, health professional and organisational users have no means of validating the quality and scientific basis of telematic services, and there is no protection for the consumer. For drugs and medical devices there are validation mechanisms, and telematic healthcare applications need similar mechanisms.

The development of telematic applications for healthcare has been very much focused on technical details and technical standards. However, the quality of medical knowledge in these systems has not been evaluated thoroughly. This has become one of the major problems in adopting medical software for clinical use. Physicians cannot rely on the software because neither they nor any other independent authority systematically evaluate it. From the industrial point of view the lack of quality assurance of medical software is also a problem. There are no tools for discriminating between high-quality products and poor products, and this is ruining the entire reputation of this industry in its infancy.

Quality and reliability of services on the Internet

The Internet has radically changed the markets of medical software. Any schoolchild is able to access medical software available through the Internet. In the near future commerce through the Internet will increase rapidly and there will be tools to charge for services directly over the net. There is already software for patient care publicly available on the Internet. To avoid legal responsibility, the authors warn that the software is not intended for clinical use. However, clinicians and patients tend to use the software for patient care. A good example is a simulator that can estimate insulin doses for patients with insulin-dependent diabetes. Clinicians, who are not allowed to use such software in their work, not infrequently inform their patients that they are able to download the software from a given Internet address to assist in insulin treatment. This approach is even more dangerous than clinicians using the software themselves, as patients do not have the necessary background knowledge to enable them to judge whether the advice provided by the software is reliable or not. Lethal consequences are therefore possible. It would also be very difficult to find out whether the software is causing an error, particularly if the physician is not aware that his or her patient is using such software.

Electronic commerce and promotion of drugs, medical devices and healthcare services on the Internet will increase rapidly in the next few years. It is highly likely that very aggressive marketing of medical services on the Internet will increase. The Internet differs from most other media in that users have difficulty in separating advertising from edited, more objective content. There are no common rules on how advertising should be presented on websites, so when information is searched for on the Internet, search engines return sites ranging from commercial material to high-quality scientific articles. This will increase the use of media to promote new drugs and medical technology. This problem was discussed recently in connection with sildenafil (Viagra®).[15]

Issues raised by patient use of the Internet

In Finland, 20% of the population uses Internet information services at least once a week. It has become a very widespread habit, especially among young patients, to look for medical information on the Internet. In addition, many elderly patients have relatives who use Internet services and look for information on their behalf. However, a number problems have already been raised by patient use of the Internet.

- Patients do not tell their physicians that they are using Internet services.
- Some companies advertise their products without any scientific evidence to support their efficacy.

The fact that patients do not tell their physicians that they are using the Internet as a source of medical information will lead to health hazards if they follow the instructions from the Internet without consulting their doctor. If adverse reactions occur, it is difficult to prove that these were caused by misleading information found on the Internet.

Some service providers advertise their products without any scientific evidence to support their use. Some drugs are claimed to be effective as sexual stimulants, or as treatment for poor memory, confusion, depression or lack of motivation, while some claim to work against ageing or to enhance mental alertness, etc. Typically, patients try to conceal their problems from their physician and relatives and try to cope by themselves by searching for help, often from the Internet.

In the USA, government policy supports the freedom and self-regulation of the Internet. In medicine, the role of the Food and Drug Administration (FDA) is crucial. In summer 1997, the FDA gave permission for drug companies and medical device manufacturers to advertise on the Internet. It believes that high-quality medical services will be most popular among users and advertisers, and that this should then boost the quality and quantity of medical websites. In our opinion, it will boost the quantity but it is difficult to believe that it will boost the quality. It seems irresponsible to pass the judgement of medical websites over to the advertisers, whilst consumers are not in a position to judge! During the past few years a number of serious problems have arisen. For example, one website was found to be promoting 'home abortion kits' and 'female self-sterilization kits'.

It is evident that there are several important aspects of the medical Internet services, including consumer protection, stimulating high-quality medical services and creating the infrastructure for a profitable information industry in the field of medicine. All of these would benefit from some kind of registration or regulation of telematics services for healthcare.

Healthcare is one of the most crowded markets in cyberspace. More and more patients are using their home computers to check on the latest healthcare news and obtain medical advice. The problem is not to do with the fact that the patients are looking for information, but to do with the quality of information that they find.[16] Retrieving medical information from the Internet clearly poses potential hazards for consumers, and patients may delay seeking medical attention. As the Internet-browsing population continues to grow, steering patients away from sources of misleading information could one day become a standard part of preventive medicine. Attempts to eliminate inaccurate information from the Internet have not been very successful, and it is impossible to try to limit the provision of medical software by the Internet. Users need to be assisted by attempting to check existing services, providing information about good-quality services and warning about the possible risks of some software. This will not eliminate the problem altogether, but it is an important way to increase patient safety.

The future model of using telematic services

In its fifth framework programme the European Union is promoting services for citizens in the user-friendly information society. Health-care is mentioned as one major field in the framework programme. Internet services are most often presented together with a large amount of advertising. Interesting services have been used to get visitors on websites to read the commercial material. Therefore many medical information services are also presented in an environment that is not giving the best possible picture of high quality.

However, the situation is changing quite rapidly. High-quality information services which do not require any marketing material on their sites are products that users want to use and will look for in the future. In healthcare in particular, most European countries have forbidden marketing of medical services to patients. Therefore, Internet services in which medical doctors use their own name should also be free of any marketing material.

The scenario of the virtual hospital seems to be the model of telematic services. Virtual hospitals consist of networks of medical doctors and healthcare providers that are able to provide medical services through the Internet. The tools consist of asynchronous text-based and/or image consultations and synchronous video-conferencing, and in the future will also include virtual reality applications. In these virtual hospitals, liability of the clinicians is clearly defined. Medical knowledge bases and information services will form the basis of the medical library of virtual hospitals. Most medical journals will be on the Internet quite soon, and the peer-review process for these journals will remain very much the same as it is now. However, there will be a large selection of other information services, decision-support software and knowledge bases that are not published in traditional ways. An important challenge for both the Internet community and medical professionals is to create quality assurance for these applications.

The European Union recognised this need when they granted funding for a European project entitled *Towards Evaluation and Certification of Telematics Services for Health (TEAC-Health)* in 1997.[17] The project will issue recommendations for the European Commission on how to promote the quality and safety of medical telematic services.

Feasibility study on rating medical Internet services in Finland

Finland is one of the first countries in which evaluation and rating of medical Internet services has been attempted in collaboration with medical doctors and the Ministry of Social Welfare and Health. The Finnish Organisation for Health Technology Assessment (FinOHTA) will start a feasibility study on the rating of medical services. The project is restricted to websites in Finnish and Swedish.

The aim is to agree on common criteria for how medical information should be presented on the Internet. Based on these criteria, a 'Code of Conduct' will be published for the content providers, and a rating will be constructed on the criteria published in the Code of Conduct. A number of medical professionals will rate websites to study the feasibility and clarity of the Code of Conduct. In order to ensure good criteria, inter-expert variation in rating should remain quite small.

After the Code of Conduct has been agreed upon, a selection of websites is rated and the content provider is informed about the rating result. In the event of problems, the webmaster is informed that certain changes are recommended. These recommendations and ratings will be confidential during this phase and will not be shown to the public. After several weeks or months the websites are checked again to see whether the suggestion given by the rating service has had any influence on the content. This study will indicate the willingness of content providers to collaborate with a third-party rating body.

The final aim is to build a database on rated medical services that fulfil the commonly agreed Code of Conduct. These pages would be allowed to use a certain label on their site to show that the site belongs to the rated services. The rating of a third party related to the site would also be made available for Internet users.

A model incorporating the Code of Conduct and registration has already been set up by the HON (Health on the Net) Foundation.[18] What is missing is the quality assurance of the content.

The model described in this paper will be tested in Finland by the HON Finland Association[19] in the near future. The impact of the rating service will ultimately be determined by the users. The aim is to teach Internet users to be critical and to demand high-quality services. If the users are critical and avoid using unrated information services, the model will have an important role in promoting high-quality health-related information services on the Internet. A proper rating by a third party will then become a necessity for medical content providers who want to survive on the market.

Conclusion

Software and telematics are increasingly prevalent support tools in healthcare, and Internet services are enabling them to be developed and used on a seamless international basis. In addition, Internet information services are increasingly being used by both clinicians and the general public. However, unlike other essential components of healthcare, such as medical devices and pharmaceuticals, telematics and Internet services are neither quality assured nor regulated, yet it is almost impossible for the individual user to evaluate them.

This seems totally counter to contemporary concepts of risk avoidance, protection of the citizen, and ethical commerce. The need for reassurance and protection through third-party mechanisms seems urgent and obvious. The TEAC-Health and FinOHTA initiatives described in this chapter are just small steps towards the identification of potential solutions to a major challenge.

References

1 Salway JG (1990) *Drug-Test Interactions Handbook* (1e). Chapman and Hall Medical, London.

2 Young DS (1995) *Effects of Drugs on Clinical Laboratory Tests* (4e). AACC Press, Washington DC.

3 Grönroos P (1997) *Medication and Laboratory. A study on computerized monitoring of drug–test and drug–drug interactions in hospital* (dissertation). Annales Universitasis Turkuensis Series D, Turku.

4 Tryding N and Roos KA (1986) *Drug Interferences and Drug Effects in Clinical Chemistry* (4e). Apoteksbolaget, Stockholm.

5 Tryding N, Tufvesson C and Sonntag O (1996) *Drug Effects in Clinical Chemistry* (7e). AB Realtryck, Stockholm.

6 Tryding N, Galteau MM, Salway JG, Breuer J, Malya PAG and Siest G (1987) Drug interferences and drug effects in clinical chemistry. Part 7. Data banks. *J Clin Chem Clin Biochem.* **25**: 191–4.

7 Grönroos P, Irjala K and Forsström JJ (1995) Coding drug effects on laboratory tests for health care information systems. *Proc Annu Symp Comput Appl Med Care.* 449–53.

8 Schoolman HM (1991) Obligations of the expert system builder: meeting the needs of the user. *MD Comput.* **8**: 316–21.

9 Hammond P, Harris AL, Das SK and Wyatt JC (1994) Safety and decision support in oncology. *Inf Med.* **33**:371–81.

10 Wyatt JC (1994) Clinical data systems. Part 3. Development and evaluation. *Lancet.* **344**: 1682–8.

11 Wyatt JC (1995) Hospital information management: the need for clinical leadership. *BMJ.* **311**: 175–8.

12 Cosgriff PS (1994) Quality assurance of medical software. *J Med Eng Technol.* **18**: 1–10.

13 Shaw R (1994) Safety-critical software and current standards initiatives. *Comput Methods Programs Biomed.* **44**: 5–22.

14 Cavalli P (1996) False negative results in Down's syndrome screening. *Lancet.* **347**: 965–6

15 Rosen RC (1998) Sildenafil: medical advance or media event? *Lancet.* **351**: 1599–600.

16 Impicciatore P, Pandolfini C, Casella N and Bonati M (1997) Reliability of health information for the public on the World Wide Web: systematic survey of advice on managing fever in children at home. *BMJ,* **314**: 1875.

17 TEAC-Health Home page: http://www.multimedica.com/teac

18 Health on the Net (HON) Foundation homepage: www.hon.ch

19 Health on the Net (HON) Finland Association homepage: www.hon.fi

CHAPTER 5

Evaluating the impact of telemedicine on health professionals and patients

Jeremy Wyatt

Introduction

One striking feature of the emerging evidence-based, wired future of healthcare[1] is the spectacular lack of reliable evidence of telemedicine's benefits for clinical practice or patient outcomes. For example, a recent systematic review[2] included over 60 randomised trials investigating the impact of decision-support systems – a much more complex technology – on clinical practice or patient outcomes. However, there is no equivalent review on telemedicine because so few trials were found (Rosemary Currell, personal communication). A number of problems do arise in telemedicine trials, and these will be discussed in this chapter, but all of them can be overcome given the motivation. Perhaps the underlying problem is that politicians and commercial interests appear to be keen to implement telemedicine regardless of the lack of objective evidence of benefit. However, taking an evidence-based view, all expensive technology should be evaluated in order to understand not only its benefits but also any risks it poses, and how to prevent them in future.[3–5]

This chapter discusses the range of evaluation approaches that are applicable to telemedicine and the need for impact studies, describes a range of contextual and other problems that arise during studies of telemedicine and how they can be addressed, and draws some general conclusions.

Evaluation approaches

The choice of the evaluation method always depends on the underlying question.[6] There are two broad evaluation perspectives, namely *formative evaluation*, in which the developer of a technology makes regular interim measurements to inform the development process, and *summative evaluation*, in which an external evaluator determines whether the developer achieved their aims. Typically, formative evaluation assesses the system's structure and function, while summative evaluation assesses function and impact.[7] Although studies of structure and function litter the telemedicine journals and conferences, only impact studies interest clinicians and health policy makers, count as 'evidence' in the wider world, and can be included in Cochrane reviews.[8]

The need for impact studies

Some writers on telemedicine evaluation have advocated that diagnostic accuracy alone should be measured.[9] However, systems with improved accuracy – or other functions such as faster data transfer – may be of no clinical benefit at all if the previous generation was already accurate or fast enough. They could even worsen the impact if, for example, the cost of equipment or networks leads to resources being withdrawn from other clinical services. Side-effects such as noise or physical bulk could intrude on the consultation, make clinicians reluctant to use them and impair the quality of communication – the reverse of what telemedicine is intended to achieve. Although we are beginning to understand which structural features of telemedicine equipment lead to improved function, we are still decades away from understanding how improved function improves the clinical consultation. Thus we cannot rely on measuring function, but we must evaluate the impact of telemedicine systems on clinical processes and patient outcomes.

Other telemedicine evaluation frameworks have argued for use of the health economic perspective[10] to analyse the different types of questions and the methods used to answer them. If the question is solvable by a cost-benefit analysis, such as 'Should we invest in this telemedicine system rather than renal services?', this makes sense. O'Rourke *et al.*[11] propose an alternative framework, suggesting that evaluators ask the following six key questions about a telemedicine system.

1 What is the system to achieve?
2 What currently happens without it?
3 If it were in place, how often would it be used?
4 For each person who uses the system, what are the benefits?
5 What are the longer-term consequences?
6 How transferable is the system?

These are important questions to inform the development of a specific telemedicine system, but they take a narrow view. For example, questions 3 and 4 imply that speculation alone may provide the answers. Rather than asking how often the system *would* be used, an objective measure of actual usage rates is preferred, and the same is true for benefit. Moreover, question 4 assumes that all effects are benefits, and fixes system users as the judge of these. It is patients and health systems who are the main claimed beneficiaries of telemedicine. Thus a better question might be, 'What is the measured impact (benefits and side-effects) of the system on users, patients and health organisations?'.

Unfortunately, some evaluators prefer the role of local auditor, describing what happened in a specific project, while making no attempt to generalise beyond the specific clinicians, organisation or system studied. However, if we assume that most evaluators wish to produce generalisable results to inform clinical and health policy, the real question becomes: 'What is the likely impact of this or a similar telemedicine system in a new site?'. Answering this question requires reliable assessment of cause and effect, which can only be achieved using the randomised controlled trial.[6,12,13]

What controls to use

One of the most difficult issues in comparative studies and clinical trials is the choice of a suitable control against which to compare the intervention, in this case a telemedicine system or service.[4]

The choice of controls depends on the exact question and the motive underlying it. For example, if the question is 'What is the impact of a store and forward teledermatology system?', one can adopt either a pragmatic or an explanatory stance.[14] The pragmatist would compare patients 'seen' by the teledermatology system with conventional dermatology referrals, who often wait for around 3–6 months for an appointment. However, any difference in clinical practice or patient outcome may have been largely due to the very different turn-around times. The explanatory evaluator would compare patients 'seen' by the teledermatology system with those for whom their GP mailed a Polaroid photograph to the dermatologist, to unravel the added benefit of the instantaneous electronic transmission of the image from the accelerated referral process.[4] This procedure has been adopted in at least one trial of teledermatology (*see* Chapter 2 of this book) and can be adapted to other telemedicine trials – for example, using a telephone call to discuss a faxed patient summary or courier-delivered video.

The remainder of this chapter will discuss the various categories of problems which arise when one is trying to make generalisable estimates of the impact of telemedicine, and will consider how each can be resolved within a realistic budget.

Contextual problems evaluating the impact of telemedicine

Evaluation is always more difficult when the technology to be evaluated takes many forms, can be applied in many different clinical contexts, and when it concerns improved information and communication rather than, say, a drug used in a certain stage of a specific disease. If the technology has many components, the context of use influences its effectiveness, and if it is changing rapidly, this also complicates evaluation.[5] Unfortunately, all of these considerations apply to telemedicine.

Multiple forms of telemedicine and contexts of use

Telemedicine can take many forms and is used in many clinical contexts,[15] ranging from an isolated email between doctor and patient[16] to a store and forward teledermatology system or multiprofessional psychiatric case discussion conducted by video-conference (*see* Figure 5.1).

The clinical context may also vary in ways that could influence the effectiveness of telemedicine – for example, with more or less severely ill patients, better qualified staff, or reserved telemedicine sessions. This profusion of telemedicine methods and clinical applications is simultaneously a tribute to the breadth of telemedicine and a reminder of how cautious we must be when generalising findings from one type of telemedicine or context of use to another. Once we have developed a broadly acceptable classification of telemedicine[15] the picture may become clearer, but until then we must remember to define carefully the exact system and context of use when reporting evaluation results.

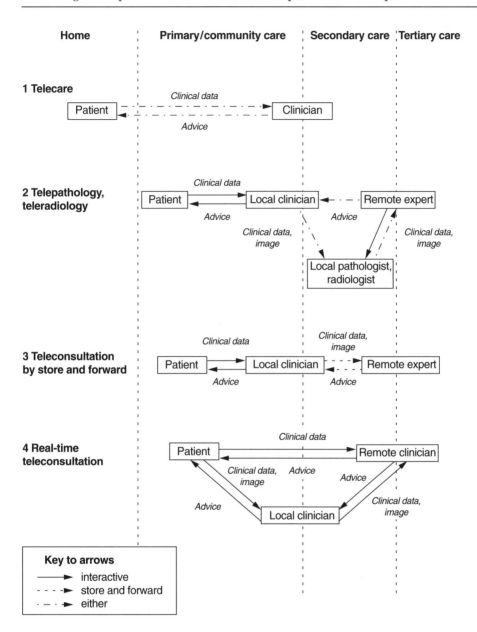

Figure 5.1: Different types of telemedicine and contexts of use.

Rapid technology change and slow learning curves

One frequent objection to summative evaluation of any information technology application in healthcare is that the system has not stabilised, and improved performance is just around the corner. A similar argument, which has retarded objective evaluation of surgical procedures and devices, is that most users are still on the learning curve.

There are several objections to these arguments. First, the claimed eventual benefits are often smaller in size and take longer to arrive than is expected by enthusiasts – or they never materialise at all. Secondly, change is not unique to information technology. New devices such as those used for coronary angioplasty offer dramatic improvements in effectiveness and lower side-effect profiles, which also always seem to be 'just around the corner'. Fortunately, there is an established culture – and regulatory requirement – to conduct rigorous studies of the impact of each generation of drugs and devices, to inform licensing decisions and clinical practice.[17] Thirdly, whatever the claimed benefits of the next generation of technology, it is the current generation which will be applied according to the results of impact studies and, given the limited investment of most health services, will stay in use for years, even when manufacturers have declared it obsolete.

Such weak arguments about rapid technology advances and doctors being on the learning curve seem to be one reason why the telemedicine field stands out from other applications of IT in healthcare for deferring objective impact studies. Fortunately, it did not deter the UK Health Technology Assessment Board from funding two large randomised trials in 1997, the results of which are already enhancing our understanding of telemedicine (*see* Chapters 2 and 7 of this book).

If the evaluator is concerned about this problem, they have several options. Apart from conducting the study as rapidly as possible, they can conduct a sensitivity analysis of critical parameters (e.g. if accuracy improved from 85% to 90%, would this have clinical implications?). Such an approach may even make it possible to avoid a study altogether. For example, a simulation of potential cost-benefits of teleradiology in Norway showed that such a service would be uneconomical.[18] Finally, it may be necessary to accept that results will become less relevant in a year or two,[17] but it is nevertheless nearly always useful to conduct rigorous studies, if only to act as a signpost marking our path through the shifting sands of health technology.

Claimed need to measure the value of telemedicine in redesigned health services

Another argument against mounting summative evaluation studies is that telemedicine's real potential can only be realised in a health service redesigned around a distributed model, with integrated information systems, staff trained in telemedicine use, etc. Any study conducted in the current health service might therefore underestimate the actual benefit that would be realised in the ideal scenario. One solution to this obstacle would be to build a very large-scale demonstrator and conduct the trial in this context, but the results may then depend heavily on the extra investment made, and may not generalise to a full-scale health service. Alternatively, evaluators can content themselves with measuring incremental changes in existing well-understood systems and cautiously model their combined results, which may either add or multiply. As long as such projections of likely benefit are subjected to external assessment and sensitivity analysis, this is the best way forward, and will certainly be cheaper and lead to more generic results than constructing a parallel health service demonstrator of sufficient size.

Problems encountered when making measurements in impact studies

An essential preliminary to conducting any demonstration study is to develop or assemble reliable tools for making the measurements.[5] However, trials of telemedicine do pose some specific measurement problems, including difficulty in blinding studies, the need for multiple measures, difficulty in measuring clinical impact, and data-collection biases.

Difficulty in blinding studies

Information technology evokes strong responses from clinicians, with some falling under its spell ('the power to mesmerise'[19]) and others objecting violently.[20] If the clinicians who use telemedicine also provide key data items about their practice or patient outcomes, they may exaggerate or underestimate the benefits according to their personal attitude to the system. The problem is made even worse if measurements are made, or extracted from clinical notes, by those who developed or introduced the system. In drug trials a number of techniques are used to blind both the clinicians managing patients and the patients themselves to whether the patient is in the study or control group. Not all methods are successful, so assessment of the adequacy of blinding is now recommended.[21]

In telemedicine trials it is difficult to blind patients or doctors to the use of telemedicine. However, it is possible to arrange for objective measures of the clinical process and patient outcome to be logged and then judged by an external assessor, who can be blinded to whether telemedicine was used or not.

Multiplicity of measurements needed

In studies of the impact of telemedicine, a wide range of measures can be used, including the following:

- attitudes of patients and staff to telemedicine and to the consultation, and perceived impact on their role (see Chapters 1 and 7 of this book)
- length of consultation and distribution of activities during the consultation (e.g. by work sampling methods)
- clinical knowledge and skills before and after exposure to telemedicine
- the accuracy and timing of clinical decisions (e.g. diagnosis or treatment choice)
- clinical actions (e.g. prescribing, test ordering and referral)
- the quality of clinical data (e.g. accuracy, completeness, timeliness)[22]
- organisational measures (e.g. length of stay or staff retention)
- costs of the telemedicine system and monetary implications of its use for patients and health services.

However, in a specific trial it is necessary to prioritise these, as making measurements is expensive, may reduce clinical efficiency and contributes to the Hawthorne effect (see below). It is also necessary to select the key measure for the sample size calculation. However, it is always sound practice to include some measures to help understand why the trial failed to show the expected benefit (if this is the case), as understanding the reasons for failure is very important.

As already mentioned, unless measurements are of objectively verifiable actions (e.g. number of laboratory tests ordered per patient), it is necessary to develop or use reliable, valid techniques (e.g. ready-made scales to measure attitudes to computers, etc.).[5,23]

Measuring impact is difficult

In some studies, surveys are used exclusively to measure the impact of systems on clinicians and patients. Although well-designed surveys can be a reliable way to assess attitudes,[24] there is a wide gulf between reported attitudes, certainty, reported actions and actual actions.[25] For this reason, the Cochrane Effective Practice and Organisational Change Group, who are reviewing the impact of methods designed to improve clinical practice, only include studies that measure clinical actions using an objective method. Such methods include review of clinical records, analysis of data collected during the consultation on special record forms, or laboratory and pharmacy records. Sometimes the patient report of whether a doctor carried out a specific procedure (e.g. counselling against cigarette smoking) may be included.

Another issue that often arises is whether to measure patient outcomes directly. Many adverse outcomes, such as surgical complications and death, are fortunately rare, so large studies would be needed to show a reduction. However, we can reliably link some clinical process measures (e.g. administration of streptokinase within 6 hours in myocardial infarction) to real outcomes. Thus studies can be much smaller when we use such validated process measures.[26]

One useful solution to the measurement problem is to exploit existing routine data systems, so long as the data collected are of high quality and address the question. This may reduce not only costs but also any Hawthorne effects, and may allow repeated measurements over time. This is important if there is concern about learning curves or the sustainability of telemedicine impact, due to decreasing usage rates.

Data-collection bias

One problem (from the evaluation perspective) that is encountered with telemedicine applications such as virtual outreach (*see* Chapter 7 of this book) is that expert involvement may lead to improved quality of patient assessment in telemedicine patients compared to controls who remain under 'GP-only' supervision. Increased identification of clinical problems in the telemedicine group may in turn lead to an apparent worsening of outcome in these patients.[4]

One solution to such data-collection bias is to ensure that all patients are 'seen' via telemedicine, except that the appointment for controls is deferred until after the GPs have made their management decisions.

Disentangling the components of telemedicine services

Telemedicine services usually add a number of extra components, such as training, reserved clinical time and accelerated appointments, to the essential advice communicated using the telemedicine system. When planning evaluations it is important to identify each potential added component and consider how it might contribute to the overall impact on patient, clinician and system (*see* Table 5.1).

Table 5.1: Some elements of a telemedicine service and ways in which each can contribute to impact

Telemedicine service component	Mode of action
Presence of novel high-technology equipment	Placebo effect – Hawthorne effect
Training and support for telemedicine service users	Hawthorne effect – see below / Training in the relevant clinical domain
Remote assistance with data collection	Check-list effect
Feedback on clinician performance	Feedback effect, contamination
Clinical advice	In-service, problem-based education, contamination
Accelerated appointments, reserved time for telemedicine cases	Intervene earlier in disease process; greater opportunity to reflect on case
Lower communication barriers	Higher referral rates, improved inter-professional rapport
Two clinicians and one patient	Changed consultation dynamics
Patient remains in home or GP setting	Changed responsibility for the patient

Some of these components can fundamentally alter the nature and dynamics of the consultation and professional roles – for example, introducing effective in-service training where none previously existed. These components are often desirable effects of introducing the telemedicine system, but they must be clearly differentiated from it, as their impact could be confounded by the impact of the advice that doctors receive via the telemedicine system. Thus care must be taken to identify added service components and ensure that their contributions to the overall impact of the telemedicine *service* are not credited to the telemedicine *system*. This is particularly important with components of the telemedicine service which vary according to local resources, clinical context, etc., as unless the relative contributions are differentiated, it will be difficult to generalise the results of the study to other settings. Thus when designing studies to quantify the impact on clinical practice or patient outcome of the clinical advice given via telemedicine – the intended mode of action – evaluators must take care to measure or control for confounding factors such as the Hawthorne, placebo, check-list and feedback effects. Some of the specific problems are discussed briefly below, and further details can be found in Chapter 7 of a recent textbook.[5]

Contamination and cluster randomised trials

Contamination occurs when the care of control patients is altered by knowledge, skills or insights acquired when the same clinicians use telemedicine for study patients. The potential result is that the care of control patients will improve to the same level as that of telemedicine patients, so reducing the apparent benefit of telemedicine, as measured by the difference between telemedicine and control patients, to zero. If the telemedicine service is one which promotes learning (e.g. by allowing generous time between patients and using experts who are inclined to teach), then contamination is likely to be more severe than if the service merely provides remote advice in a way that cannot be emulated locally.

The solution to contamination is to conduct a cluster randomised trial,[27] randomising doctors, practices or even hospitals[25] to use of the telemedicine system. Whereas one might study 400–800 patients in a patient-randomised trial, a typical cluster-randomised trial will include around 20–30 randomised clinicians or practices and 30–100 patients per practice, giving a total in the range 600–3000. Such studies need specific sample-size calculations which take into account the intra-cluster correlation coefficient; often they need 20–60% more patients.[28] As well as its effect on the required sample size, the cluster design also has implications for the analysis of study results.[29]

Training in the relevant clinical domain

Training staff to use the telemedicine system usually means that they also spend time learning about the specific clinical problems, acquiring up-to-date knowledge and skills from an expert even before the telemedicine system goes live. This may explain why, in some studies of telemedicine, the use of the telemedicine consultation tool itself has rapidly reduced (Richard Wootton, personal communication). In a trial in which clinicians rather than patients are randomised, it also means that the impact of the telemedicine system may be exaggerated, as some of the true impact is due to the special training that the telemedicine users received.

The solution is to provide similar clinical skills training in the subject for control staff, or to study a 'training-only group'. As staff may object to being trained without ever gaining access to telemedicine, evaluators can arrange for the telemedicine training to be followed by a gap of a few weeks before the telemedicine service is made available, during which time the impact of the training alone can be assessed.

Hawthorne effect

It is common knowledge, and has been shown in psychological experiments, that our performance improves when we know that we are being studied.[30] Thus in telemedicine trials where staff are randomised we would expect a greater increase in performance in the telemedicine group due to the continuing presence of telemedicine equipment, while control staff have no such daily reminder of the trial. The intensity of data-collection activities can also influence the size of this effect.

In order to minimise the Hawthorne effect, evaluators should exploit routine data sources as much as possible and monitor performance over time, as the Hawthorne effect tends to fade over a few weeks. An alternative is to balance the Hawthorne effect by comparing the impact of two different telemedicine systems, or to use more complex methods, such as the balanced incomplete block design.[5]

Check-list effect

Clinical performance (e.g. accuracy of diagnosis) is improved when doctors use structured data-collection methods.[31] This effect is greater in junior doctors working outside their area of expertise. In telemedicine trials, the expert may impose a standard data-collection routine that itself improves performance, even before any advice is given. It would then be a mistake to attribute such performance improvements to the advice, or to the telemedicine link, as the standard list of data to be collected could be made available to all remote sites by low-technology methods.

One solution which balances the check-list effect on both groups is to agree on standard data-collection forms for use in both control and telemedicine patients/practices. Alternatively, a third 'data-collection only' group could be recruited. This could use a paper form, or it could conceivably be telemedicine assisted but receive no advice.

Accelerated, reserved appointments for patients

Often a telemedicine system is introduced with other changes to the service, such as more rapid appointments or even reserved telemedicine clinics which may be more leisurely than ordinary clinics. This in turn allows more time for clinical reflection and may lead to disease processes being arrested at an earlier stage in their natural history. Some of the resulting improvements in patient care or outcomes are thus not due to the telemedicine system but to the accompanying changes in service delivery.

The solution is to arrange similar accelerated but physical appointments in a clinic with a similar workload for control patients,[4] as in the randomised trial of virtual outreach described by Wallace (*see* Chapter 7 of this book) or to recruit a third group of patients who are seen using telemedicine but after the normal out-patient waiting time, in the normal rushed clinic.

Summary and conclusions

Telemedicine is a significant innovation with great potential to improve healthcare[16,32] but, like all powerful innovations, its impact (both beneficial and harmful) needs to be assessed reliably. Often telemedicine behaves as a complex intervention, resembling a service more than an IT system, making it difficult to disentangle the various elements and a mistake to attribute all benefits to the telemedicine system. This complexity does introduce a number of biases which need to be considered when designing studies. It also makes it more difficult to configure a successful telemedicine service, which is why we should take pains to publish not only successful studies but also the lessons learned from negative ones. A registry of telemedicine trials would considerably assist those conducting systematic reviews in this area, and is a well-established strategy for advancing evidence-based healthcare.[33]

Negative studies reveal our current limited understanding of clinical communication and information needs[34,35] and highlight the need for more studies to inform the configuration of telemedicine systems.[36,37] They also highlight the need to define precisely the study context and specific information deficits in order to allow all telemedicine study results – positive or negative – to be interpreted correctly.

Health-service managers and policy-makers should be aware of the unknown cost-benefit of telemedicine applications, and defer wider implementation of telemedicine until the results of current studies are known. It would be preferable, while we are still so short of reliable evidence on telemedicine, to encourage clinicians to consider entering all patients into well-designed randomised trials, although we would stop short of placing them in the National Centre for Clinical Excellence's category C, which is only suitable for NHS use in the context of a well-designed study.[38]

This chapter has described some of the methodological details that need to be addressed when designing reliable studies of telemedicine, and we would suggest that telemedicine developers should seek the advice of those with skills in health technology assessment, medical statistics, health economics and other appropriate specialist fields when evaluating such systems.

In conclusion, we would warn those approached with a view to implementing any specific telemedicine application to look for reliable evidence, to beware of trials driven by technology and funded by developers,[4] and to remember the key question 'Telemedicine: interesting or necessary?' (with apologies to the authors of a previously published article[39]).

References

1 Wyatt JC and Keen JR (1998) The NHS's new information strategy (editorial). *BMJ*. **317**: 900.

2 Hunt DL, Haynes RB, Hanna SE and Smith K (1998) Effects of computer-based clinical decision support systems on physician performance and patient outcomes: a systematic review. *JAMA*. **280**: 1339–46.

3 Jennett B (1988) Assessment of clinical technologies. *Int J Technol Assess Health Care*. **4**: 435–45.

4 Wyatt JC (1996) Telemedicine trials: clinical pull or technology push (commentary)? *BMJ*. **313**: 1380–1.

5 Friedman C and Wyatt J (1997, reprinted 1998) *Evaluation Methods in Medical Informatics*. Springer-Verlag, New York.

6 Sackett DL and Wennberg JE (1997) Choosing the best research design for each question. *BMJ*. **315**: 1636.

7 Wyatt J (1997) Quantitative evaluation of clinical software, exemplified by decision support systems. *Int J Med Informatics*. **47**: 165–73.

8 Cochrane Library (1999) Effective practice and organisation of care: Cochrane review group methods. In: *The Cochrane Library. Issue 2*. Update Software, Oxford.

9 Grigsby J, Schlenker RE, Kaehny MM, Shaughessy PW and Sandberg EJ (1995) Analytic framework for evaluation of telemedicine. *Telemedicine J*. **1**: 31–9.

10 Bashshur RL (1995) On the definition and evaluation of telemedicine. *Telemedicine J*. **1**: 19–30.

11 O'Rourke CM and Gallivan S (1997) *Telemedicine: evaluation or stagnation?* CORU Discussion paper 443. Clinical Operational Research Unit, University College, London.

12   Sacks H, Chalmers TC and Smith H (1982) Randomised vs. historical controls for clinical trials. *Am J Med.* **72**: 233–40.

13   Julious SA and Mullee MA (1994) Confounding and Simpson's paradox. *BMJ.* **309**: 1480–1.

14   Schwartz D and Lellouch J (1967) Explanatory and pragmatic attitudes in therapeutic trials. *J Chronic Dis.* **20**: 637–48.

15   Wallace S, Wyatt J and Taylor P (1998) Telemedicine in the NHS for the millennium and beyond. *Postgrad Med J.* **74**: 721–8.

16   Borowitz S and Wyatt J (1998) The origin, content and workload of electronic mail consultations. *JAMA.* **280**: 1321–4.

17   Goodman C (1996) The moving target problem and other lessons from percutaneous transluminal coronary angioplasty. In: A Szczepura and J Kankaanpaa (eds) *Assessment of Health Care Technologies.* John Wiley, Chichester, 123–40.

18   Halvorsen PA and Kristiansen IS (1996) Radiology services for remote communities. *BMJ.* **312**: 1333–6.

19   Maclaren P and Ball CJ (1995) Telemedicine: lessons remain unheeded. *BMJ.* **310**: 1390–1.

20   Sears-Williams L (1992) Microchips vs. stethoscopes: Calgary hospital MDs face off over controversial computer system. *Can Med Assoc J.* **147**: 1534–47.

21   Begg C, Cho M, Eastwood S *et al.* (1996) Improving the quality of reporting of randomised controlled trials: the CONSORT statement. *JAMA.* **276**: 637–9.

22   Wyatt JC and Wright P (1998) Medical records 1: Design should help use of patient data. *Lancet.* **352**: 1375–8.

23   Streiner DL and Norman GR (1995) *Health Measurement Scales.* Oxford University Press, Oxford.

24   Oppenheim AN (1991) *Questionnaire Design, Interviewing and Attitude Measurement.* Pinter Publishers, London.

25   Wyatt J, Paterson-Brown S, Johanson R, Altman DG, Bradburn M and Fisk N (1998) Trial of outreach visits to enhance use of systematic reviews in 25 obstetric units. *BMJ.* **317**: 1041–6.

26   Mant J and Hicks N (1995) Detecting differences in quality of care: the sensitivity of measures of process and outcome in treating acute myocardial infarction. *BMJ.* **311**: 793–6.

27   Altman DG and Bland JM (1998) Units of analysis. *BMJ.* **314**: 1874.

28   Thompson SG, Pyke SD and Hary RJ (1997) The design and analysis of paired cluster randomised trials: an application of meta-analysis techniques. *Stat Med.* **16**: 2063–79.

29   Divine GW, Brown JT and Frazier LM (1992) The unit of analysis error in studies about physicians' patient care behaviour. *J Gen Intern Med.* **7**: 623–9.

30   Roethligsburger FJ and Dickson WJ (1939) *Management and the Worker.* Harvard University Press, Cambridge, MD.

31   Adams ID, Chan M, Clifford PC *et al.* (1986) Computer-aided diagnosis of acute abdominal pain: a multicentre study. *BMJ.* **293**: 800–4.

32   Wyatt JC (1996) Advances in communication and information technology: implications for health services. In: M Peckham and R Smith (eds) *The Scientific Basis of Health Services.* BMJ Publishing, London, 124–37.

33   Dickersin K (1992) Why register clinical trials? – revisited. *Control Clin Trials.* **13**: 170–7.

34   Wyatt J (1991) Use and sources of medical knowledge. *Lancet.* **338**: 1368–73.

35   Smith R (1996) What clinical information do doctors need? *BMJ.* **313**: 1062–8.

36 Coiera E (1998) Four myths about the information revolution in health care. In: J Lenaghan (ed) *Rethinking IT and Health.* Institute for Public Policy Research, London, 16–29.

37 Wyatt J (1998) Four barriers to realising the information revolution in health care. In: J Lenaghan (ed) *Rethinking IT and Health.* Institute for Public Policy Research, London, 100–22.

38 Rawlins M (1999) In pursuit of quality: the National Institute for Clinical Excellence. *Lancet.* **353**: 1079–82.

39 Miller PL and Sittig DF (1990) The evaluation of clinical decision support systems: what is necessary versus what is interesting. *Med Inf.* **15**: 185–90.

CHAPTER 6

Telemedicine and the protection of patient interests: experiences from Norway

Irma Iversen

In the years to come the use of telemedicine will increase. Accordingly, we need to look into the corresponding legal issues, and the new arrangements relating to primary care and health institutions. As a patient ombudsman I shall take the opportunity in this chapter to discuss the legal issues on behalf of those patients who require treatment by a specialist.

The Norwegian health system and the patient ombudsman

Norway has 19 counties in total, of which 17 counties have a patient ombudsman. These counties are grouped into five health regions. The counties own the hospitals with the exception of the National Hospital, which is owned by the government.

Akershus county, where I have worked as a patient ombudsman for the last 10 years, is the second largest county, surrounding Oslo city with about 468 000 inhabitants. The county has several hospitals, and the Norwegian National Hospital in Oslo is the regional hospital of the county.

The Norwegian government has granted patients with severe illness the right to be treated within 3 months (emergencies are of course treated immediately). Akershus County belongs to Region II, which includes seven counties. It is almost impossible for a patient in our county to cross the border of the health region to obtain treatment in another region. Consequently, the patient needs to be guided and advised if he or she does not wish to be treated in the nearest hospital. The general practitioners know their local hospital and the specialists working there, and therefore mostly tend to use that hospital, although they do not have to do so.

Today most of the patients in Norway have to be treated in a hospital owned by the municipal county in which they live. The county has both the duty and the right to deliver the health service in a hospital. Therefore, in a sense, the disease of the patients is 'owned' by the county! However, if the county cannot take care of the patients in due time, the patients in Akershus County can be treated in a hospital outside the county, and Akershus County is then responsible for the expenses. By contrast, if a patient is treated in a private hospital and complications occur, then there is a major problem of who will pay the financial compensation.

Legal aspects in using telemedicine

Three main situations may occur when using telemedicine.

1 The patient or the county may choose to use telemedicine provided by the hospitals owned by Akershus County.
2 The county hospital needs to be advised by a specialist in the National Hospital.
3 The patient is treated in a private hospital, which needs advice from outside.

Health practice by remote experts creates problems with regard to the responsibility for and 'ownership' of the treatment and its result. If the patient is treated by means of telemedicine by one specialist in a county hospital and by another physician performing the treatment in another county hospital, there will be no legal problem. However, if the specialist is located at the National Hospital and the treatment is given to a patient in a county hospital, there could be a problem in dealing with the two different owners of these hospitals. This potential problem has been largely solved in Norway by making it the duty of these hospitals to use the same insurance free of charge for the patients. However, the problem will arise if the specialist is located at the National Hospital and the patient is treated in a private hospital. The requirement to have insurance for non-private hospitals started in Norway by agreement from 1 January 1988, in Sweden from 1 July 1991 and in Denmark from 1 July 1992.

Telemedicine provides opportunities for both diagnosis and treatment, and therefore the legal issues have to be addressed. In 1992,

Norway introduced a new law to give all patients the same rights for compensation for complications without looking into the ownership of the hospitals (Compensation for Complications: Norwegian Official Report (NOU) 1992: 6). We now have to obtain agreement between the county and the National Hospital, and between the county and the private hospitals. The Norwegian government delivered a report to the Storting (parliament) in 1999, in the form of a 'White Paper'. It is likely that the law about the duty of insurance for all our specialists and hospitals will be passed in June 2000, thus solving the legal problems concerning responsibility and compensation.

Data protection and responsibility for the medical record

General protection for each patient is primarily associated with handling of information about decision making so that the health workers can give the patient the correct treatment. However, we have not yet solved the problem of data protection and responsibility for the medical record. Accordingly, in telemedicine all individual decisions about the treatment of a particular patient have to be made in relation to both to the actual diagnosis of the patient and the right of each patient to decide on the type of treatment. Thus the patient being treated by telemedicine has to agree to the different diagnostic and treatment procedures.

The interests of the patient consist of the following components:

- user-friendly administration
- the right of the patient to choose
- confidentiality and security
- completeness – the patient's right to know the data and to correct it
- protection against misuse.

User-friendly administration

Before telemedicine is accepted as a routine procedure in the health system, the patient has to give their consent. This will result in both an extended obligation of the general practitioners (GPs) to inform the patient when telemedicine is to be used, and a closer contact between the GP and the specialist than currently prevails. The GP must have the primary responsibility to inform the patient about the different legal aspects of telemedicine and the option of choosing between the various specialists working in the field of telemedicine.

The right of the patient to choose

The right of the patient to decide is stated in the Norwegian law for physicians, paragraph 25. This particular law will be replaced in January 2001 by implementation of the Law of Health Workers No. 64, which passed the Storting (Parliament) in July 1999.

In 1999 we also passed our first law, known as the law of patient rights, where the patient's rights to take part in the decision on treatment are written in paragraph 3–1. According to this paragraph the patient can also choose between the different diagnostic and therapeutic procedures available. In another paragraph (paragraph 3–2) it is stated that the patient must be given the necessary information in order to understand their health condition and the nature of the available treatment options.

Confidentiality and security

When the patient has made a decision and given consent to the use of telemedicine, the question of confidentiality has to be resolved. The patient must be informed that strangers might listen to the conversation, but that the equipment is as secure as possible. The patient also has to be informed that the health workers have no responsibility or accountability for the technical equipment. This last responsibility is regulated in a law about telecommunication that was enacted on 1 July 1998. This law (paragraph 9–3) deals with the security aspects. The owner of the telecommunication network is required to prevent this information from being made available to others. The health workers' duty to protect their information about the patient is regulated in the law mentioned above with regard to the health workers (paragraph 21).

Completeness – the patient's right to know the data and to correct it

Telemedicine in fact alters the relationship from one between the patient and one doctor to an interaction with several doctors and other health workers at the same time.

The consultations, diagnosis and treatment decisions need to be documented. The demand for completeness in telemedicine may include the use of video and other tape recordings. However, the patient must give their consent to the use of such recordings.

The patient has the right to look at the medical records and all other information about him or herself, including tape recordings. It must be decided which person in the diagnostic and treatment chain is responsible for documenting and storing the information obtained. The patient must also be informed and told who is responsible for the documentation, so that they can have access to the information.

The law about the patient's rights (paragraph 5–2) gives the patient the right to demand that the information in the medical record is corrected or deleted. The current regulation (IK-18/97) gives the patient the following rights to correct the medical record:

- when objective faults exist, such as wrong spelling of their name or date of birth
- when a medical record exists without a patient–physician relationship
- when the recorded information is irrelevant to the treatment and is found to be offensive by the patient.

If the physician fails to correct the medical record, the patient can complain to the medical officer appointed by the government in the county. If the medical officer also refuses to make the desired corrections, the Data Directorate should be contacted in order to obtain their decision.

Protection against misuse

The information in the medical record can only be used for the purpose for which it was collected, or for other reasons specified by law. However, the patient can give consent to the physician and other health workers for it to be used for research purposes. To avoid misuse of the data stored on various tapes, the tapes have to be deleted after a specified period of time.

Conclusion

In the future telemedicine will become an important part of medicine, and will involve co-operation with other countries with which Norway has medical agreements. According to the new law about patient rights, every county must establish a patient ombudsman, who will then be required to inform the patient about their rights, which include the option of choosing between different hospitals in the entire country.

The need for information for patients is increasing, and I have also observed this in my work as a patient ombudsman. In 1989 my office received 800 telephone consultations, but this number had increased to 3600 in 1999. Issues of telemedicine will add another dimension to this need for advice.

CHAPTER 7

Telemedicine in primary care: evaluating the effects on health practice and health practitioners

Paul Wallace

Introduction

Primary care is the provision of integrated accessible healthcare services by clinicians who are accountable for addressing the large majority of healthcare needs, developing a sustained partnership with patients and practising in the context of family and community. In the UK there are many components of primary healthcare provision, but the principal one is general practice. General practitioners operate as independent contractors within the National Health Service, each general practitioner having on average some 2000 registered patients. There has been an increased tendency over the last 40 years for general practitioners (GPs) to work in groups, and well over 90% of GPs now operate in group practices, with an average of three to five partners looking after a total of around 10 000 patients. Such group practices usually employ additional staff, including practice nurses, receptionists and a practice manager, and there is an increasing tendency for other professionals to be employed, such as counsellors and complementary practitioners. There are further primary care providers who, like general practitioners, operate as independent contractors within the NHS, including opticians, pharmacists and dentists. An additional important area of primary care is currently provided through community trusts, who are responsible for providing services delivered by district nurses, health visitors, chiropodists, community psychiatric nurses, physiotherapists and others. With such a complex framework of primary care provision on a highly devolved basis, communication problems and networking are particularly difficult.

Although telemedicine is likely to be of major importance for secondary and tertiary healthcare providers, it is potentially particularly important for primary care, where practitioners tend to work in isolation, dispersed in a range of settings within the community and often with little regular contact either with each other or with the secondary and tertiary services. Health telematics thus has the potential to make a substantial impact within this healthcare sector.

There are a number of factors which are likely to facilitate such developments. These include the advent of extensive computerisation within general practice, with at least 95% of practices now being computerised.[1]

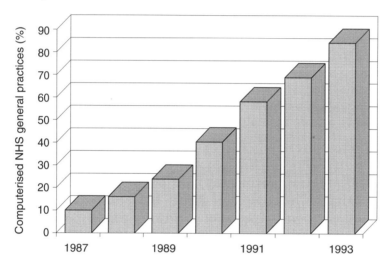

Figure 7.1: Computerisation in general practice in the UK (after Fry).[1]

The NHS information strategy document and the advent of the NHS network will provide further support for the implementation of tele-medicine within primary care.[2] In addition, the tendency during the last decade to shift the provision of care increasingly from the secondary and tertiary sectors to the primary care sector looks set to continue. Furthermore, primary care may well take advantage of telemedicine in developing the commissioning role envisaged by the establishment of Primary Care Groups and Primary Care Trusts. Although not all practices are making maximum use of computers, and there are still major practical and medico-legal problems relating to the provision of practices with telematic links, there is none the less enormous potential for the rapid development of health telematic applications in primary care.

Some applications of telemedicine in primary care

There are a number of areas within primary care where telemedicine is likely to make a potentially important contribution to changing health practice and health practitioners. These include the following:

- transmission of clinical records
- on-line access to patient information systems
- store and forward advice/referral (with or without image)
- medical monitoring in home-care setting
- networking for educational purposes
- teleconferenced medical consultation at the interface between primary and secondary care.

Many of these applications are already being developed both within the UK and elsewhere. Electronic records are now almost ubiquitous within general practice, although there remain substantial problems of standardisation both between practices (when patients re-register) and between practices and hospital and other agencies. On-line access to patient information systems has been developed in the USA where, for example, interactive programmes relating to treatment of conditions such as benign prostatic hypertrophy and carcinoma of the breast have been developed and tested. This type of application is now being tested in the UK. A number of projects in the UK and elsewhere have tested the effectiveness of using store and forward technology. This approach has been used successfully in dermatology to enable specialists to make decisions, based on transmitted images, about the likely urgency of treatment for pigmented lesions.[3] Medical monitoring in home-care settings has been explored in relation to elderly patients living in residential care and nursing homes, as well as patients in 'hospital at home' settings.[4,5] Networking for educational purposes may be of special significance in primary care, where it is often difficult for GPs and nurses to find time to travel to educational meetings. The TeleEducation And Medicine (TEAM) project in Wales explored this application and found that it was especially helpful for practice-based nurses.[6]

The likely impact of these telematics applications on health practice and health practitioners will depend on the way in which they are introduced, and the degree to which they address and provide real solutions to the problems of everyday work within the service. Such questions are crucially important for policy development, and need to be addressed by thorough evaluation of cost-effectiveness. The evaluations should use rigorous scientific methodologies, and the results should be presented in such a way as to assist policy-makers in making informed decisions about implementation.

The virtual outreach project: evaluating telemedicine at the interface between primary and secondary care

The virtual outreach project on teleconferenced joint medical consultations is an example of the development of a structured approach to the assessment of the potential impact of one telemedicine application on health practice and health practitioners in both the primary and secondary care sectors.

Development of the project

Joint teleconferenced consultations make use of video-conferencing technology to enable GPs, patients and hospital consultants to discuss jointly problems which would otherwise be referred for review in a routine hospital out-patients clinic or in a specialist outreach clinic in general practice. The technological support for this application of telemedicine comes from the use of ISDN 2 lines with a PC-based video-conferencing application. The project has been developed by staff at the Royal Free Hopital School of Medicine and University College Medical School, London, together with colleagues at the Centre for Health Informatics in Swansea and the Institute for Rural Health in mid-Wales. The idea of using telemedicine to facilitate joint consultations arose from an awareness on the part of both clinicians and patients of the frequent communication problems which impair the quality of out-patient referral, and from work in The Netherlands involving joint consultations between GPs, orthopaedic surgeons and patients.[7] A randomised controlled trial demonstrated that patients seen jointly in this way underwent fewer investigations, received fewer medical interventions and were less likely to need out-patient follow-up than those who were seen in a routine out-patients clinic. Furthermore, levels of well-being and satisfaction among the patients were high, and there was evidence of an important learning effect on the clinicians. The use of video-conferencing to enable such joint consultations to take place in the GP surgery with the specialist 'beamed in' via the video-link seemed to have much to offer. In London, a project was developed by the Department of Primary Care and Population Sciences at the Royal Free Hospital School of Medicine. Initially this was a feasibility trial, supported by the Camden and Islington Health Authority and British Telecom. Following the success of the feasibility study, a grant was secured from the NHS Research and Development Primary/Secondary Interface Programme to conduct a pilot study for a randomised controlled trial. The pilot and feasibility study involved linking general practices to hospital-based specialists in a number of specialty areas including urology, gastroenterology, paediatrics, endocrinology, dermatology, orthopaedics, oncology, ENT medicine and rheumatology.

In Wales, the TEAM project was established – using a similar technology – to develop and evaluate the potential benefits of video-conferencing in the area of dermatology.

Evaluation results

Although at this stage the two projects were not closely co-ordinated, the evaluation was of a similar nature and produced broadly equivalent results. Within the Royal Free project, evaluation concentrated on obtaining the views of providers (both hospital clinicians and GPs) and patients about the quality of the consultation.[8] The majority of the GPs and consultants taking part in the project were positive about the level of communication and about the ability to establish a rapport with the patient. There were some reservations about the technical quality of the teleconsults amongst consultants, some of whom were unhappy about the quality of sound and images. The responses of the patients were in general very positive, with well over 90% stating that their experience was either satisfactory or very satisfactory, and the large majority indicated that they would be happy to use this form of consultation in the future.

Some examples of comments include the following.

- 'This is an excellent breakthrough for children, who may be frightened of hospitals.'
- 'A very positive experience. Please make it more widely available.'
- 'Having my own doctor present made the whole thing much easier to deal with and a very useful result was achieved.'
- 'When I go to hospital as an out-patient, I wait at least 1 to 2 hours to see a doctor. After the long wait I forget all the questions I want to ask.'

In London, the pilot study went on to test the potential to conduct a randomised controlled trial comparing virtual outreach with routine hospital out-patient consultations. A cluster randomised trial design was adopted, and a series of standard instruments were used to evaluate patient satisfaction and well-being in both groups.[9] In addition, a methodology was developed to enable an economic evaluation to be undertaken. The pilot study demonstrated the feasibility and value of adopting this form of structured approach to assessment. Furthermore, despite the small sample size, it demonstrated strong evidence of enhanced patient satisfaction with virtual outreach as compared to routine out-patients appointments.

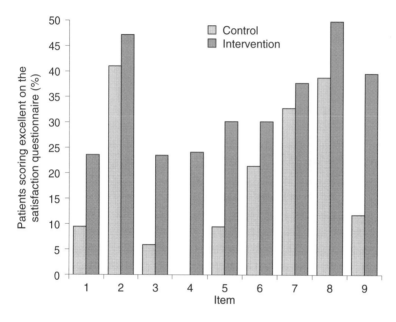

Questionnaire items
1 How long you waited to get an appointment
2 Convenience of the location of the hospital/surgery
3 Getting through to the surgery/hospital by telephone
4 Length of time spent waiting at the surgery/hospital
5 Time spent in the consultation
6 Explanation of what was done for you
7 The technical skills (thoroughness, carefulness, competence) of the doctor(s) you saw
8 The personal manner (courtesy, respect, sensitivity, friendliness) of the doctor(s) you saw
9 The overall visit

Figure 7.2: Patient satisfaction with teleconferenced medical consultations vs. routine out-patient appointments (after Harrison *et al.*).[9]

On the basis of the above findings, an application for a full-scale randomised controlled trial to be conducted in Wales and London was successfully submitted to the NHS Research and Development Health Technology Assessment Programme. This trial is based at the Royal Free Hospital in London and the Royal Shropshire Hospital Trust, and involves more than 50 GPs at each of the sites, together with some 20 hospital consultants.

Changes to clinical culture

The virtual outreach study not only shows how a systematic approach to evaluation can be adopted, but also gives indications about how telemedicine may impact on health practice and health practitioners in the primary care sector. It suggests that there may well be advantages to bringing services to patients rather than patients going to services, but it also showed that the use of the new technology produced a new social situation in which there were no clear rules of engagement. The nature of the joint teleconferenced consultations produced changes in the power relationship for all parties, with loss of 'hands-on' experience for both the patient and the consultant, and greater exposure of the professionals to one another. There were organisational issues relating to the increased demands for punctuality, and a dependence on unfamiliar technology. Although these considerations related specifically to the virtual outreach project, it is likely that they will be more broadly applicable across a range of telemedicine applications. The likelihood is that telemedicine, in whatever form it develops within the NHS, will create and facilitate organisational change, with accompanying changes in social relationships and networks between professionals.

Ensuring appropriateness

Innovation of this kind is likely to be perceived as threatening, and implementation strategies will need to take account of this. If they are to prove successful, they will need to address the problems of scepticism, behavioural inertia and 'technophobia'. Appropriate and reliable market research techniques will need to be employed to ensure that applications are developed to suit the real needs of the patients and healthcare providers. High-quality product design will be needed to address the issues of technophobia, and a properly structured approach to assessment, including rigorous economic evaluation, should be adopted in order to produce robust evidence to inform plans for implementation.

References

1   Fry J (1993) *General Practice: the facts.* Radcliffe Medical Press, Oxford.
2   Department of Health (1998) *Information and Management Technology Strategy.* Department of Health, London.
3   Loane MA, Gore HE, Corbett R *et al.* (1997) Effect of camera performance on diagnostic accuracy; preliminary report from the Northern Ireland arm of the UK Multicentre Teledermatology Trial. *J Telemed Telecare.* **3**: 83–8.
4   Allen A, Roman L, Cox R and Cardwell B (1996) Home health visits using a cable television network: user satisfaction. *J Telemed Telecare.* **2**: 92–4.
5   Wootton R, Loane M, Mair F *et al.* (1998) A joint US–UK study of home telenursing. *J Telemed Telecare.* **4**: 83–5.
6   Freeman K, Wynn Jones J, Groves-Phillips S and Lewis L (1996) *Teleconsulting: a practical account of pitfalls, problems and promise. Experience from the TEAM project group.* Proceedings of the 'Telemed 95' Conference, London, November 1995.
7   Vierhout WPM, Knottnerus JA, van Ooij A *et al.* (1995) Effectiveness of joint consultation sessions of general practitioners and orthopaedic surgeons for locomotor system disorders. *Lancet.* **346**: 990– 4.
8   Harrison R, Clayton W and Wallace P (1996) Can telemedicine be used to improve communication between primary and secondary care? *BMJ.* **313**: 1377–81.
9   Harrison R, Clayton W and Wallace P (1999) Cluster randomised controlled trial of virtual outreach – a pilot study. *J Telemed Telecare.* **2** :126–30.

CHAPTER 8

Ensuring clinical ownership of telemedicine: a nursing case study

*Lorraine Gerrard, Adrian M Grant and
J Ross Maclean*

Introduction

Rapid technical developments in video and telecommunication techno-logies since the early 1990s have led to a resurgence of interest in telemedicine. However, the uptake of interactive video-conferencing as a tool by which to deliver healthcare has been relatively modest, despite extravagant claims by its proponents (primarily those from the informa-tion technology and telecommunications industries), although this situation may be set to change. The government White Paper, *The New NHS: Modern, Dependable*, embraced plans to 'develop telemedi-cine to ensure specialist skills are available to all parts of the country'.[1] However, past experience suggests caution is needed. Time and time again telemedicine services have been implemented enthusiastically, only to fail to mature into routine clinical services subsequently. Of the many factors implicated, the main reason appears to be lack of consideration of the human resource implications.[2] Provision of the telecommunication facility alone is not sufficient, and telemedicine services are likely to be ineffective or fail unless the implications for skill mix and human resources are addressed.[3,4]

There are two broad models for live, interactive teleconsulting. The first is doctor-to-specialist consultation (doctor-centred), and the second is nurse-to-specialist consultation (nurse-centred). Although doctor-centred services may be particularly appropriate for occasional complex clinical problems, this model appears to be relatively imprac-tical and resource intensive for routine cases, as it requires two doctors to be available at the same time. In contrast, nurse-centred services seem to be more realistic and viable. For example, a large proportion of telemedicine consultations are likely to be for non-life threatening routine or even urgent referrals, for which nurses are increasingly providing initial care.

Nurses could play a pivotal role in telemedicine services, which would be implicitly linked with broader changes in the roles of nurses working in acute and community services, for which the government is also committed to provide encouragement and support.[1] We explored reactions to nurse-led telemedicine and their possible implications in a project that formed part of the Department of Health (England and Wales) Human Resource and Effectiveness Research Initiative.

Methods

The study was divided into two parts. The first part was a broad systematic review of the literature addressing human resource issues for healthcare professionals involved in telemedicine. Although this consisted predominantly of technical reports and feasibility studies of doctor-centred telemedicine services, relevant findings were used to inform the research questions in the second part. This was a case study of all four UK sites which involved nurses in a teleconsultative role. Two sites were established services, while telemedicine was no longer operational in the other two. The four sites consisted of an Accident and Emergency Service in the north-east of Scotland between Peterhead and Aberdeen, a Minor Accident and Treatment Service in Wembley linked to the Central Middlesex Hospital in London, tele-dermatology and tele-education links between general practices in rural central Wales, and the Minor Treatment Centres in South Westminster and Parsons Green in London, linked with an Accident and Emergency Department in Belfast. A total of 36 individuals directly involved in these services were interviewed using semi-structured schedules, and they included nurses ($n=25$), general practitioners ($n=4$), advising medical consultants ($n=2$), 'service managers' ($n=2$), and research personnel ($n=2$). All interviews were conducted after the implementation of the service. The selection of individuals represented a combination of theoretical sampling and opportunistic choices made within the time constraints of each site visit. The interview transcripts were analysed. Data-coding was the first step in generating themes by linking raw data and theory. 'Open strategies' derived from the research questions were revised and specified as data collection proceeded. New topics were developed which became codes and were added to the coding theme, and all discrepancies were checked. Cross-checking, unfreezing and reconfiguring persisted throughout this iterative cycle of field work and analysis.

Results

We developed a framework – the project 'model' (Figure 8.1) – to reflect a logical chain of evidence. Each factor within the model was independently emphasised by different informants with different healthcare roles at each field site. Causal links were deduced directly and indirectly from the field study and literature. The logical coherence, grounded in the iterative analysis, was verified in each successive wave of data collection within the cross-site analysis. The initial tentative map of the main factors was refined and modified until it was complete, and the stream from antecedents to outcomes stood alone within the two overall themes of procedural (organisational) and professional (individual) change. The model is grounded within and specific to this project, although future generalisations could be made with regard to new technologies and nurses. The key findings are described below using the refined framework of the model.

Procedural issues relating to the procedures and practicalities of telemedicine services

Introduction of concept

Nurses reported that their involvement in initial decisions to introduce telemedicine services had been minimal. They mentioned that they were often unclear about the purpose of the proposed application and how their role was likely to be affected, and some reported that they were confused by what they later saw as distorted claims made by those with a vested interest in promoting the equipment. Some nurses were also uncomfortable with the use of the word 'telemedicine' for what they saw as nurse-centred services.

> '*I'm sure downgrading it would make it less frightening. It needs to be viewed as less high tech so that it becomes a normal part of daily living, so you can use it without thinking, like a washing machine. At the moment it's a bit scary because if it goes wrong, or if you press the wrong thing, you don't know what to do. . . . If it could be a teaching tool for us, right at the start of our nursing, it would be fine, accepted. We'd be as confident with it as my kids are with computers.*'
>
> (Nurse)

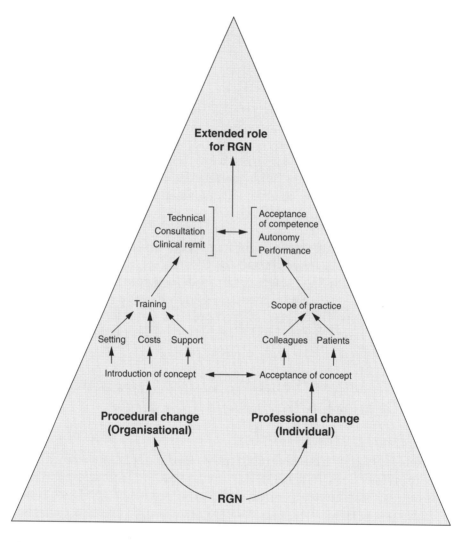

Figure 8.1: Model illustrating the human resource implications for the nursing profession in developing telemedicine within the NHS.

Setting

The services were established in pre-existing buildings, and for this reason some nurses felt that the positioning of the equipment was sometimes unsatisfactory. Respondents felt that this influenced the extent to which the equipment was used, and nurses in some centres considered that this aspect should have received more attention before the service was introduced.

Costs and resources

Nurses worried about the 'responsibility' of using what they perceived to be expensive equipment and the possibility of damaging it. Many of them felt ill prepared, saying that the resources allocated for staff training had been inadequate. In two sites, nurses felt disheartened because they considered that the telemedicine equipment would be withdrawn on completion of the project regardless of whether or not the service was successful.

Support

Although this varied between centres, some nurses reported having felt isolated and unsupported, and wanting more recognition from colleagues and managers, with more effective back-up from the system vendor/product provider. Technical support manuals and 'trouble-shooting' guides were often unavailable or 'incomprehensible'. There was a perceived lack of central co-ordination, and more senior respondents argued for a broader perspective to telemedicine planning to include local, regional and national policy-makers and managers.

Training – technical

Nurses reported initial fears about the technology, which were sometimes debilitating. There were marked differences in their previous keyboard skills, such that baseline training needs varied widely. There were complaints that the training was over-technical and didactic, with insufficient opportunities for 'hands-on' experience. In some places, the introduction of nurses to telemedicine was complicated by technical teething problems that would have been avoided if there had been thorough field-testing before any attempt was made to use the system in clinical practice.

Training – consultation

A commonly expressed view was that teleconsultations take longer and that this is not recognised, for example, by appropriate clinic scheduling. The nurses also often felt self-conscious about seeing themselves on screen, which they again linked with inadequate prior training.

> 'You might feel a bit intimidated by it initially, 'cause you think "Oh my God, does my hair really look like that, my make-up and

*don't I look fat!" Now you go in and use it, you don't bat an eyelid,
so you do get used to it.'*

(Nurse)

Training – clinical remit

Nurses confirmed that access to telemedicine was associated with an
extension of their clinical roles. Respondents recognised that this
required additional training, irrespective of telemedicine. Two specific
areas that were identified were when to seek assistance over the video
link, and the use of medical terminology when discussing clinical cases
over the telemedicine link.

Professional issues relating to the impact of the technology on the nurses and their workplace

Acceptance of concept

A perceived poor definition of the nurses' extended roles was associated
by some respondents with both inappropriate training and lack of
acceptance and misunderstandings among other health professionals.
Uncertainty about these issues was also reflected in wide variation in
nursing titles and training at the four sites.

Colleagues

Although some nursing and other health professional colleagues were
reported to be very supportive, others were not. Respondents linked this
to the extent of colleagues' involvement in the introduction of the
service, and perceptions of encroachment of the nurses' enhanced role
(for example, the ordering and initial interpretation of X-rays).

Patients

All of the nurses reported a supportive response from patients in
relation to both telemedicine and the nurses' expanded role. The
nurses described how patients perceived the new services as 'technology being shaped to meet their needs'. The advantages for patients
were considered to be improved access to specialist care, increased
quality of care through specialist consultations, and reduced transporta-
tion problems and concerns.

'They're asking if they can use it, it's not just us who miss it, the patients really thought it was wonderful.'

(Nurse)

Scope of practice

Respondents regarded the increasing scope of nursing practice as the main advantage of nurse-centred telemedicine. They saw telemedicine as an excellent learning medium which did not appear to cause de-skilling. Varying consent and record-keeping procedures had been found, with some respondents perceiving a need for standardisation.

'No you're not deskilled at all, it adds to your skills. In a way it increases your autonomy 'cause the patient doesn't have to go anywhere else, and you learn from the link, from the new information that's given to you. So your skills and your knowledge increase all the time and you become a bit more independent of other medical professionals.'

(Nurse)

Acceptance of competence

With regard to nurses' competence to fulfil their expanded role, doctors in the four centres carried vicarious liability provided that the nurses followed both the United Kingdom Central Council for Nursing, Midwifery and Health Visiting (UKCC)[5,6] and local guidelines.

'I don't think there is a ceiling (to the extended role), as long as you are aware of your accountability. . . If we weren't prepared to defend what we're doing, and if we didn't feel that what we were doing was safe, and we were putting the patients in danger, then we wouldn't do anything.'

(Nurse)

Autonomy

Although telemedicine was recognised as being simply a means of delivering healthcare services, its introduction appears to have had wide implications. The nurses' enhanced role was popular, but it had a significant impact on intra- and inter-professional working relationships. Nurses generally felt empowered by their medical colleagues and local managers to take on an extended role, and considered that they had greater autonomy through telemedicine, but this was not always the case.

> 'I think that while telemedicine is here, the patient gets a holistic approach to care, they don't have to go anywhere else, and I think that's what makes it a winner.'

> (Nurse)

Performance

In two centres, the performance of the nurses was generally guided and limited by sets of protocols.[7] This was reported to work well. There had been no medico-legal problems at any of the four sites.

> 'There's more autonomy [with telemedicine]. We can see almost anything that comes through the door within our protocols, although the severity varies. When I first came here as a nurse practitioner, I found the need to use it was much more than it is now. Maybe that's because I'm actually more experienced now, and don't actually have to ask for advice as much as I did in the beginning. But the point is that it's there if we ever need to use it, it's a safety net both for us and the patients. If we hadn't had telemedicine to fall back on I don't think we'd have come this far. I don't think we'd have had the confidence to see the things we've seen. It means we can have a second opinion at the flick of a switch really.'

> (Nurse)

Discussion

Our study was based on only four sites, and the choice of interviewees was partly opportunistic, which may limit its generalisability. Ideally we should now study a larger number of nurse-led services and compare and contrast those that have failed with those that have succeeded. Nevertheless, there do appear to be clear messages for the introduction of new telemedicine services, particularly those that are nurse centred.

Feelings of 'ownership' were important, and success appeared to be more likely if all of the groups affected were involved from as early a stage as possible. This illustrates a more general issue with regard to the introduction of new technology into healthcare and the importance of ownership.[8,9] Those who were interviewed argued that the practical implications of a new service should be considered before its introduction. These include the physical location of the equipment (which may need a new building), development of standard procedures and protocols, and adjustment to scheduling. Thorough equipment testing on site before its use in training and practice would probably have resulted in some of the problems described being avoided. Respondents felt that it would have been better if it had been presented to new users as (another) tool for healthcare, avoiding extravagant claims about what it could achieve. Another recurring issue was that of good-quality ongoing technical support, which should always be in place and include easily understood manuals for minor problem solving.

Initial training varied widely between the four centres and was not always judged to be adequate and appropriate. Based on what we were told, it should include technical information, so that there is a basic understanding of how the system works, as well as meaningful practical 'hands-on' experience. The training should also reflect the changes in role that go with a new service, such as responsibility for procedures, and the new levels of communication with other health professionals. This intensity of training has resource implications that should be built in from the start, especially as the more thorough the training, the more satisfied the nurses were found to be. It was clear from what we were told that ongoing support and training are also important, and standards which monitor and review this aspect should accompany evolving services.

Most important of all is clarity of the purpose of a telemedicine service, and hence of professional roles and responsibilities. We consider that the time has come to look at telemedicine services more broadly in order to develop minimum standards for training, support and common procedures. Telemedicine has retained a stage of technical development such that new services should only be established if there is an expectation that they will be integrated into routine healthcare (provided that piloting is successful).

Given the limitations of our study, more research is needed to determine how patients and health professionals respond to telemedicine. The starting point for telemedicine applications should be the identification of the needs and preferences of consumers and providers, taking a user-driven rather than a technology-driven perspective.

We believe that the factors which the study has identified should be addressed if telehealth services, especially those involving nurses, are to be implemented. Users' opinions should be systematically gathered both before equipment is acquired and while it is being implemented. Specifically, our model (Figure 8.1) provides a framework for considering these human resource implications, and if it is used we believe that the main causes of failure can be addressed.

It is not easy to get the human components of telemedicine (both individuals and organisations) to work well together and with a complex technology that is still evolving. Our study has shown that all of the personnel involved must be prepared for rapid learning that is implicit in the implementation of change. The challenge is to maximise the potential for the primary operator to use new technology. With regard to nurse-centred services, the nurse must not be viewed as a technician, nor should telemedicine be seen as a substitute for doctor availability. To achieve this balance, the nurse must become a core member of the team, determining and planning the implementation of such healthcare systems. This is likely to enhance job satisfaction, improve job performance (including patient care) and lead to a more effective and efficient use of this evolving technology.

References

1 Department of Health (1997) *The New NHS: modern, dependable*. The Stationery Office, London.
2 Allen A and Perednia D (1996) Telemedicine and the health care executive. *Telemed Today*. **Winter**: 4–9, 22–3.
3 Maclean JR (1996) Telemedicine and the nurse: the benefit or burden of new technology? *J Telemed Telecare*. **2**: 54–6.
4 Scott JC and Neuberger NI (1996) Human factors and the acceptance of telemedicine. In: MJ Field (ed) *Telemedicine: a guide to assessing telecommunications in healthcare*. Institute of Medicine, National Academy of Sciences, National Academy Press, Washington DC.
5 United Kingdom Central Council for Nursing, Midwifery and Health Visiting (UKCC) (1992) *Code of Professional Conduct for the Nurse, Midwife and Health Visitor*. UKCC, London.
6 United Kingdom Central Council for Nursing, Midwifery and Health Visiting (UKCC) (1992) *The Scope of Professional Practice*. UKCC, London.
7 Tachakra S, Sivakumar A, Hayes J and Dawood M (1997) A protocol for telemedical consultation. *J Telemed Telecare*. **3**: 63–168.
8 Przestrzelski D (1987) Decentralization. Are nurses satisfied? *J Nursing Admin*. **17**: 23–8.
9 Jillson-Boostrom I (1990) *Worker participation in technology assessment: medical advances and the changing roles of nurses* (PhD thesis). Polytechnic of Central London, London.

CHAPTER 9

Societal and professional issues of telematics in healthcare

José Garcia de Ancos

Introduction

Telematics in healthcare can be defined as the delivery of health services and/or the exchange of health-related information using electronic media. The government's White Paper, *Information for Health*, supports the use in the NHS of technologies such as telemedicine and telecare to improve quality of care, and to prevent unnecessary travelling for patients and referrals to hospitals.[1] In order for medical practice to benefit from the use of information and communication technologies, these should (1) be adapted to the wider *information needs* of the clinicians, (2) take account of the *economic context* of the health sector in which they are applied and (3) be integrated in *society* at levels where information might be more likely to produce health benefits. Telematics applications are converging with the Internet as a medium for global communication, and telematics will be of value to health professionals if it can deliver valuable information within this environment. Its economic advantages have been promoted for about 10 years, but unless these are backed up by hard evidence, the claims of cost-effectiveness will remain unproven.[2] Clinical need, not selective computer literacy in society, should be the main priority driving the spread of health-related information developments. This is important in order to avoid information-related inequalities or – to use the common information jargon – '*e*-inequalities' which are emerging among those who for financial, medical, cultural, language or other reasons might be excluded from the potential benefits of information in healthcare. It is within this broader social context that this chapter will examine the following areas of 'information' in healthcare:

1 the potential benefits for patients and doctors
2 its impact on the patient–doctor relationship
3 its integration in clinical practice.

Benefits of information for patients and doctors

Doctors cannot practise medicine effectively without up-to-date, reliable information. In turn, patients increasingly want to be the doctor's partner either as a participant in the management of their condition, or in order to check that their doctor is offering the best treatment possible. However, both doctors and patients may not always have access to all of the information that they want. It is here where the 'information gaps' could in theory be filled by information solutions supporting telematics technologies such as access to an electronic patient record, a telemedicine link or a telecare service. Some of these investments might provide a cost-effective alternative to other forms of healthcare.[3] Explicit partnerships between doctors and patients as a direct result of developments in health-related information might improve both examination and quality of care of certain medical conditions. Unfortunately, there have been at least three obstacles to this route in healthcare.

First, some information systems have been developed in the past without addressing the information needs of clinicians. Instead they have been dominated by medical informatics concerns with technological solutions which may lack a clinical justification for their existence.[4] Evidence to date on the clinical benefits of some telemedicine initiatives is patchy.[5] Communication between primary and secondary care might be enhanced by projects with clear methodologies and specific outcomes. However, telemedicine consultations take longer than face-to-face encounters, and they carry with them training, ethical and quality burdens which have workload implications for doctors. If for a particular condition the alternatives to telemedicine are simpler, one could well question its value. All of these factors may affect the workload ceilings of telemedicine in clinical practice.[6]

Secondly, whilst investment in information is both desirable and inevitable for any health service that wants to manage its resources adequately, it is a dauntingly complex and expensive enterprise. The government's White Paper, *Information for Health*, states the intention to invest in developing an electronic NHS. However, it is difficult to estimate the resource need as many of the indirect costs will arise during the implementation phases. Education and training of staff, the intangible costs associated with upgrading of systems and telecommunication networks, the unification of standards of communication, and the implementation of security and confidentiality guidelines will all have to be taken into account. Until now, telemedicine has remained marginal to NHS business despite previous attempts to gain a substantial niche in healthcare in the UK.[7] Whether or not this situation will change soon and telematics will play a major role in the future NHS will depend on the successful implementation of all of the above issues.

Thirdly, information should be subject to appropriate peer review and evaluation processes. In the same way that randomised controlled trials are appropriate to inform the evidence base of some healthcare interventions, the evaluation processes of information should integrate the cost-effectiveness, quality and ethical aspects of its impact in healthcare.[8] Most health systems make some of their decisions on economic grounds.[9,10] If, at present, information and communications technologies are sufficiently developed to become an integral part of the health service, they should be compared with the alternatives. If the public has easier access to health-related information via the Internet, it is important to assess whether certain populations might be disadvantaged as a result of this. If these changes in society, education[11] and the practice of medicine are creating a revolution in healthcare,[12] it is important to ensure that they will have a positive impact on the socio-economically disadvantaged sectors of society.

The debate on information as an investment can be considered part of priority setting in health systems. Paradoxically, society and individuals may have more information about more interventions and treatments via information technology, and yet may be unable to benefit from them. It is important for society to have an explicit debate on the health service that they want.[13]

Investing in the development of a comprehensive electronic health record for every citizen by the year 2005 is certainly an important priority for the NHS at present. However, given the choice, the public might prefer, for example, to have more highly trained nurses or better doctors. It remains unclear how investments in a whole panoply of telematics services could help to avoid more ill health than, say, adequate public health measures to prevent the high morbidity, mortality and social costs of, say, tobacco smoking or traffic road accidents.

Failure to identify the right beneficiaries of information investments might also lead to failure to integrate information with clinical practice.[14] People, patients, doctors, nurses and others have been trained to play new roles as a result of changes in the delivery of healthcare. The use of electronic media must be supported by evidence that it has been chosen in order to provide an improved healthcare system.

Medical informatics integrated with telecommunication applications such as telematics will no doubt play a role in many health systems. It is important for health professionals to take responsibility to ensure that the information that emerges from these applications will primarily be of help to patients. Weighing the potentially conflicting loyalties between technology, limited resources and individual needs of patients might well pose some difficulties for doctors.[15]

Impact of information on the patient–doctor relationship

The increasing use of information sources such as e-mail[16] and the World Wide Web,[17] and the use of the Internet as a medium for communication provides both opportunities for an improved exchange of information between patients and doctors and cause for concern in the healthcare sector.[18]

The capacity of e-mail to combine text linked to other sources of information makes it a very effective method of communication, with obvious applications in a healthcare setting.[16] However, clinical practice realities of ethics, confidentiality and quality of care limit its applicability in doctor–patient interactions.[19] E-mail cannot and should not be a substitute for a clinical examination, as it cannot by itself discern which messages may require urgent medical attention, and it may therefore waste valuable time.[20] Some consider e-mail to be a form of 'telemedicine', with its own educational, legal and ethical repercussions for clinicians. Secure transmission of information is not a safeguard for its proper use in clinical practice. Informed consent and confidentiality are part of the patient–doctor relationship and the clinical record. E-mail is part of both aspects, and therefore both constraints apply.[21]

The Internet is one of the most popular methods of obtaining health-related information either from sources such as Medline or directly from health-related sites on the Web. More than a third of all users of Medline are members of the general public, with another third from the medical profession and the remainder belonging to research and other institutions. More than half of all the information made available via the Web has a health-related content. This is a promising situation which makes the Internet and the Web central to a more open relationship between patients and health professionals. This effect will be largely determined by who has access to the vast quantities of information available, and by establishing reliable methods to enable users to scrutinise the quality of that information.

When planning the development and implementation of Web-based interactive interfaces between health systems and patients in the fields of telematics and medical informatics, it is important to bear in mind the principles of access, high quality and relevance to patients.[22] Doctors are a major source of information, and by taking communications and information technologies seriously they may promote improved patient–doctor relationships. Free access to high-quality medical knowledge on the Internet is perhaps a more useful way to counterbalance misleading information than attempts to control its content.[23] However, there has been little general evaluation of whether accessing poor-quality information on the Web may be compensated for by an equal ability of the user to access more high-quality information.[24]

Barriers to the integration of information in clinical practice

Medical informatics should be of help to clinical practitioners in those areas where computers can replace routine monitoring tasks or aid the management of specific conditions. Unfortunately these unquestionable advantages may be limited if medical informatics develops clinical support systems which fail to reassure clinicians about issues of quality and standards of practice. These standards of practice need to be accepted by clinicians and reflect the acceptable evidence basis of clinical interventions. If doctors perceive clinical support systems to be a 'controlling' mechanism rather than a clinical practice tool, or if they doubt the evidence basis of their recommendations, they may not trust a computer to make decisions about the management of their patients. Some of these drawbacks might be avoided by tailoring the development of clinical support systems to the local needs of clinicians. This flexibility could include standards of care which apply locally, and a degree of ownership in the development, maintenance, and evaluation of its content.[25]

Other barriers that need to be removed are far from new, and have to do with telecommunications infrastructure and adequate software. A health service developed around the use of electronic communications should be able to deliver better information about patients and a more accurate and detailed clinical service using innovative technology. The updating of health-sector telecommunications infrastructures and the adoption of integrated software is still a constraint between healthcare providers. Outside the NHS similar situations exist. Many households in the UK are either not connected to on-line information resources, or the users of such facilities may lack easy access to relevant information resources. Cultural, language and literacy problems are all important elements of the equation of technological, economic and communications issues affecting access to electronically stored information.[26]

The 'quality' problems of the Internet as a resource for medical knowledge have been tackled by trying to address the 'quantity' problem. This seems to be a reasonable approach, as it is probably easier to produce poor-quality health-related information more cheaply on the Web. The delivery, i.e. the Web packaging, may yield a well-presented product but not necessarily a high-quality one. However, regulation of the supply side of information on the Internet is difficult and open to debate. Attempts have been made by, for example, the World Health Organization and many others to develop quality criteria to assess the standard of health-related information on the Internet. Some of these rating instruments cannot easily define what they are measuring, but evaluations including variables such as content, functions, impact on patient outcomes and peer review might be of value.[27] These measures may be further improved by measures that enable the user to filter information more effectively.[28,29] Although this is difficult, certain key criteria seem to be emerging from various sources, so we may be a step closer to a simple set of criteria that users can understand and use.[30]

Co-operation between government and clinicians can benefit the implementation of confidentiality and security policies which are relevant to the needs of those working in clinical practice.[31] A better understanding of the clinical need for certain information flows across healthcare institutions would help to increase privacy, and might contribute to defining the rights of access to information networks in the health sector.

However, for all of these policies to work there is also a need for training and awareness locally, at the level of the individual health professional entering or accessing data from a computer terminal. None of the efforts to make the communication of information in the NHS more effectively secure, confidential and private can be achieved unless there is a political commitment at the highest levels to ensure that the future electronic NHS will comply with these training and educational agendas for clinicians.

A way forward for patients, doctors, industry and government?

The educational, cultural and training agendas of health professionals, government and the general public are as important as the choice of adequate sources of information and communication technologies in healthcare.

Users of information need to be empowered with tools which will enable clinicians and patients to discriminate better between the beneficial and potentially harmful applications in terms of clinical value, quality, ethics and confidentiality.

The cost-effectiveness of using information applications such as telematics needs to be evaluated further, as does the need to invest in solutions which may have a medical informatics rather than a clinical objective in mind.

Confidentiality, security and privacy of data transmission limit the potential value of information technology applications in healthcare. The involvement of health professionals in developing information which is appropriate for a given clinical purpose is a core principle of good clinical practice.

Limitations on the extent to which information can be integrated in a healthcare setting will be reduced as some of the issues discussed in this paper are gradually resolved. Those using information and communication technologies in medical practice should fight scepticism, be explicit about its merits, and promote it as a clinically justified option for their patients.

References

1 NHS Executive (1998) *Information for Health.* NHS Executive, Leeds.
2 Anon (1995) Telemedicine: fad or future (editorial). *Lancet.* **345**: 73–5.
3 Fisk NM (1996) Fetal telemedicine: six month pilot of real-time ultrasound and video consultation between the Isle of Wight and London. *Br J Obstet Gynaecol.* **103**: 1092–5.
4 Smith R (1996) What clinical information do doctors need? *BMJ.* **313**: 1062–8.
5 McLaren P (1995) Telemedicine: lessons remain unheeded. *BMJ.* **310**: 1390–1.
6 Harrison R, Clayton W and Wallace P (1996) Can telemedicine be used to improve communication between primary and secondary care? *BMJ.* **313**: 1377–81.
7 Wooton R (1998) Telemedicine in the National Health Service. *J R Soc Med.* **91**: 614–21.
8 Heathfield H (1998) Evaluating information technology in health care: barriers and challenges. *BMJ.* **316**: 1959–61.
9 Robinson R (1993) What does it mean? *BMJ.* **307**: 670–3.
10 Udvarhelyi S (1992) Cost-effectiveness and cost-benefit analysis in the medical literature. Are the methods being used correctly? *Ann Intern Med.* **116** :238–44.
11 Davis D (1998) Global health, global learning. *BMJ.* **316**: 385–9.
12 Towle A (1998) Changes in health care and continuing medical education for the 21st century. *BMJ.* **316**: 301–4.
13 Holm S (1998) Goodbye to the simple solutions: the second phase of priority setting in health care. *BMJ.* **317**: 1000–2.

14 Keen J (1998) Rethinking NHS networking. *BMJ*. **316**: 1291–3.

15 Sabin J (1998) Fairness as a problem of love and the heart: a clinician's perspective on priority setting. *BMJ*. **317**: 1002–4.

16 Borowitz SM (1998) The origin, content and workload of e-mail consultations. *JAMA*. **280**: 1321–4.

17 Donald A (1998) Medicine and health on the Internet. The good, the bad and the ugly. *JAMA*. **280**: 1303–4.

18 Silberg W (1997) Assessing, controlling, and assuring the quality of medical information on the internet. *JAMA*.**277**: 1244–5.

19 Eysenbach G (1998) Responses to unsolicited patient e-mail requests for medical advice on the world wide web. *JAMA*. **280** :1333–5.

20 Rutter T (1998) Doctors warn of dangers of the internet. *BMJ*. **317**: 1103.

21 Spielberg AR (1998) On call and online, sociohistorical, legal and ethical implications of e-mail for the patient–physician relationship. *JAMA*. **280**: 1353–9.

22 Ferguson T (1998) Digital doctoring: opportunities and challenges in electronic patient physician communications (editorial). *JAMA*. **280**: 1361–2.

23 Eng T (1998) Access to health information and support. *JAMA*. **280**: 1371–4.

24 Coiera E (1997) Information, epidemics and the immunity of the Internet. *BMJ*. **317**: 1469–70.

25 Classen DC (1998) Clinical decision support systems to improve clinical practice and quality of care (editorial). *JAMA*. **280**: 1360–1.

26 Mikta M (1998) Developing countries find telemedicine forges links to more care and research. *JAMA*. **280**: 1295–6.

27 Bingham C (1998) The *Medical Journal of Australia* Internet peer review study. *Lancet*. **352**: 441–5.

28 Wyatt J (1997) Commentary: measuring quality and impact of the World Wide Web. *BMJ*. **314**: 1879–81.

29 Eysenbach G (1998) Towards quality management of medical information on the Internet: evaluation, labelling and filtering of information. *BMJ*. **316** :1496–500.

30 Kim P (1999) Published criteria for evaluating health-related sites. *BMJ*. **318**: 647–9.

31 The Caldicott Committee (1997) *Report on the Review of Patient-Identifiable Information*. Department of Health, London.

CHAPTER 10

Clinical issues: controls and safeguards

Stephanie Bown

Telemedicine has the potential to revolutionise the delivery of health-care but, despite the fundamental rethink that its integration into mainstream medicine will demand, it will still be subject to current law and medical ethics, and these are unlikely to change significantly.

The NHS Estates Health Guidance Note on telemedicine[1] states:

'Ultimately telemedicine is a vehicle for the delivery of health and as such people have the same rights to quality of care and clinicians owe the same duty of care and have the same interprofessional relationships, as with the conventional delivery of health.'

Most of the medico-legal issues raised by telemedicine can be accommodated by established legal and ethical principles. However, because telecommunications can so easily cut across regional and national boundaries, new and undecided questions have arisen with regard to jurisdiction.

Jurisdictional issues

Provisions in the Civil Jurisdiction and Judgements Acts of 1982 and 1991 allow a plaintiff to issue proceedings in the defendant's country of domicile if both plaintiff and defendant are citizens of the EU; this is known as the primary jurisdiction. Alternatively, the case may be brought in the place where the harmful event occurred, which has been interpreted by the European Court of Justice to include both the place where the negligent advice was given and the place where the resultant harm was suffered.

Thus if a German patient on holiday in Spain teleconsults with a doctor in England and suffers harm because of poor advice given by the doctor, the doctor's liability may be determined according to German,

Spanish or English law. The jurisdiction issues may be further complicated by a product liability claim against the manufacturers or dis-
tributors of faulty equipment.

Because issues of jurisdiction world-wide are not clear, all doctors
practising telemedicine across international boundaries should be
aware of the risk of negligence claims being brought against them in
places other than their country of residence. There may also be problems with the legal aspect of effectively practising medicine in a
country in which one is not registered or licensed to practise. Those
who are indemnified by one of the UK-based medical protection
organisations should be particularly cautious about engaging in telemedicine that may give rise to claims in the USA or Canada, as both of
these jurisdictions are specifically excluded from their indemnity
provision.

The provider of general information

For many years doctors and other healthcare workers have provided
general medical and healthcare information via books, papers, journals,
radio and television for both professional and lay audiences. Contributing information to a non-interactive website is simply another medium
for this type of communication.

Doctors who advertise telemedicine services (on the Internet or
elsewhere) should follow the General Medical Council's detailed guidance on advertising.[2] If they are associated with a telemedical company
or organisation, doctors should also be careful to avoid including
information that is to their own professional advantage in its promo-
tional literature.

The doctor–patient relationship

The principles of consent and confidentiality apply to teleconsultations
as much as to conventional consultations. There are inevitable confidentiality problems posed by telemedical practice, but patients have a
right to expect that the security of on-line medical databanks can be
assured.

Confidentiality

Data and images transmitted over electronic networks are an extension
of the medical record, and failure to protect them properly may give rise

to both criminal and civil liabilities as well as accountability before the General Medical Council in respect of serious professional misconduct.

Examples of the kind of breaches of confidentiality that can occur include leaving telemedical data on a visible or accessible screen, forwarding information electronically to an inappropriate party, inadequate protection of records from computer hackers, and alteration of medical records.

All of these possibilities and more must be anticipated and policies developed to prevent them from occurring.

Medical records

Doctors should document details of each telemedical consultation just as they would for a face-to-face encounter. This is not only essential for the purpose of good continuity of care, but it may also serve as real evidence of what transpired in the consultation should this later be brought into question.

Access to and the processing of documents arising from teleconsultations are covered by the 1990 Access to Health Records Act and the 1998 Data Protection Act. The first gives the patient the right to request a copy of the documents, and the second controls the way in which the documents are held and used.

Consent

If innovative techniques or treatments are being employed, extra care should be taken to ensure that the patient fully understands the risks, benefits and limitations of the technology. Otherwise, the patient concerned will not be in a position to give valid consent. Once the express consent of the patient has been obtained, this should be documented.

If, during a teleconsultation, another person is present 'off-screen', the patient should be made aware of this situation and consent to it. Similarly, they should be told if the consultation is being recorded and their consent should be obtained for any use to which the recording may be put, such as training or teaching. If the patient withholds their consent, this wish must be respected and documented. The General Medical Council has published specific guidance on audio and video recordings.[3]

During any consultation a doctor must put him- or herself in a position to make a sound clinical judgement or to make alternative arrangements to ensure that someone else will be in a position to exercise that judgement. In a teleconsultation, the doctor has no access to non-visual senses such as touch and smell. Poor clarity of a transmitted image may result in negligent treatment. These risks can be reduced if the doctor recognises the limits placed on diagnostic ability by the quality of information available in the teleconsultation setting.

The learned telespecialist

When a doctor seeks advice via the Internet from a specialist in a particular field, this is akin to a referral and the specialist takes on certain responsibilities. The scope of those responsibilities and the individuals to whom they are owed has not been specifically established in the context of telemedicine, but the general scope is well established in case law.

It can be argued that the specialist is only providing services to the referring doctor, and not to the patient. The specialist has no opportunity to examine the patient fully, and would not be in a position to carry out treatment him- or herself. He or she merely advises the referring doctor, who then has to decide whether to act on the advice, and so the referring physician retains responsibility with the patient for the acts of the specialist.

An alternative and, to my mind, stronger argument is that the teleconsultation establishes a doctor–patient relationship akin to that of a conventional consultation, and the specialist has a duty to the patient to exercise his or her judgement to a standard recognised as proper by a reasonable, responsible and rational body of specialists in the same field. The fact that the specialist has no physical contact with the patient does not destroy the consultative nature of the interaction – consider the position of radiologists and pathologists.

In distinguishing between these two propositions, it is important to consider whether the patient is aware of the local doctor's contact with the distant telespecialist.

The learned intermediary

Before referring a case to a teleconsultant, the learned intermediary should be sure that the consultant has the necessary credentials. He or she must provide the teleconsultant with sufficient information, including all relevant symptoms, background medical information and physical findings, for the specialist to make an informed judgement.

He or she must also communicate with the patient about the use of the telespecialist and the respective roles of the specialist and the intermediary. He or she carries the burden of explaining the process and obtaining the patient's consent.

In turn, the specialist should provide the learned intermediary with enough clear information, preferably in written form, to allow the intermediary to make a diagnosis or treatment plan based on the consultation.

Lack of effective communication is one of the commonest factors contributing to adverse incidents leading to negligence claims.

Risk management

Telemedicine raises some new problems and expands the boundaries of others. Doctors practising telemedicine should give some thought to the following points.

Consider:
- who wants your opinion and why
- to whom you owe a duty of care
- the limitations of technology on the quality and quantity of information available to the teleconsultant.

Implement or introduce:
- policies to protect the security and confidentiality of medical data
- protocols for live teleconsultations to prevent communication problems
- specific equipment guidelines
- training in telemedicine
- audit, and adverse incident management.

Conclusion

The use of telemedicine will grow as its applications increase and barriers decrease. Nothing to date has indicated that the application of telemedicine is unusually hazardous, but there are some specific risks that can be guarded against by following established legal and ethical principles and striving to improve communication. Some jurisdictional issues have yet to be clarified.

References

1 NHS Estates (1997) *Telemedicine: health guidance note*. The Stationery Office, London.
2 General Medical Council (1995) *Advertising*. General Medical Council, London.
3 General Medical Council (1997) *Making and Using Visual and Audio Recordings of Patients*. General Medical Council, London.

CHAPTER 11

Preserving privacy in telemedicine: putting the patient first

Jonathan Bamford

Debates may rage over how best to define 'telemedicine' in a clear and succinct manner. Is it 'medicine practised at a distance'[1] or is it 'the use of telecommunications and information technology to provide health-care services to persons at a distance from the provider'[2]? Whatever definition is settled upon, there is one common feature – the lifeblood of the systems is information, usually personal information and often sensitive personal information.

The deployment of telemedicine applications reflects a common challenge and one that is not limited to the healthcare sector. How does the implementation of increasingly powerful, information-hungry technology proceed alongside the need to respect the rights of the individuals whose personal and often sensitive information populates such systems? Undoubtedly the introduction of telemedicine into the healthcare arena could, if properly deployed, represent a major step forward. It promises prompt and timely care with attendant efficiency savings for the hard-pressed healthcare budget. However, the possibil-ities afforded by technology can be so beguiling to the healthcare professional and administrator that they may lose sight of the person it is intended to serve, namely the individual patient.

Data protection legislation

In many countries around the world the need to balance the possibil-ities offered by new technologies with the fundamental human rights of the citizen has manifested itself in the introduction of data protection legislation. Such legislation sets standards by which the data users must operate, and legal safeguards for the individual data subject.

Within the European Union, member states have adopted the EU Data Protection Directive (95/46/EC). This recognises that data-processing systems are designed to serve mankind and that they must respect the fundamental rights and freedoms of individuals, most notably the right to privacy. The Directive reinforces Article 8 of the European Convention for the Protection of Fundamental Rights and Freedoms, which requires respect for the private life of individuals, and the Directive seeks to ensure that an equivalently high level of protection exists throughout the European Union.

Member states are implementing legislation either for the first time or to upgrade their existing legislation to the Directive's requirements. Within the UK the Data Protection Act 1998 has reached the statute book and will replace the existing Data Protection Act 1984 by 24 October 2001. At the same time, the UK Parliament has passed the Human Rights Act 1998, which assimilates the principal elements of the European Convention for the Protection of Fundamental Rights and Freedoms into UK law. This includes the right in Article 8 to respect for private and family life. There has never been a clearer requirement for those handling information about individuals to develop practices which respect the privacy rights of individuals.

The professional ethics context

Those working within the healthcare system are already required to operate to the highest level of professional ethics, and are therefore well versed in the need to respect and maintain patient confidentiality. This is also one important element of compliance with data protection standards, but there are many others which need to be incorporated into any telemedical applications before they are deployed. Some of them are obvious, such as security precautions, while others are less so, such as the need to ensure that only the minimum information is held and that retention periods are set and adhered to when processing data. Experience has taught data users that failure to incorporate the necessary safeguards at the system and work-flow design stage often results in expensive consequences (both financial and human). There is also a tendency for those with a limited understanding of data-protection legislation to assume that such legislation only applies to information held on computers. The UK's new legislation is much wider, covering a whole range of automatically processed personal information, such as video images, as well as extending to certain manual records. The same legally enforceable standards will apply to all such personal information.

The political policy context

The impetus for the deployment of new technologies is not solely restricted to those within the healthcare system, as political initiatives are also providing a spur to such developments. The UK government has recently published a White Paper entitled *Modernising Government*.[3] This has a wide agenda ranging from the modernisation of the democratic framework of the country through to improvements in the delivery of public services. The government's desire to have common public services available 24 hours a day and 7 days a week manifests itself in the extension of NHS Direct to the whole country by the end of 2000. Although still in its trial phase, NHS Direct permits round-the-clock healthcare advice to be made available to individuals by the use of telephone 'call-centre' technology. At present advice is not provided on the basis of access to individual patient records, but as time goes by and electronic patient records are developed, this will become inevitable if such a scheme is to fulfil its potential. All of the challenges that arise for those providing similar services, such as telephone banking, including the difficulties of verifying the caller's identity and restricting disclosure to the correct individual, will arise. The simple concept that advice can be given over the telephone will need to be modified to include appropriate identity verification safeguards if increasingly substantive personal data are to be consulted, interchanged and recorded. In its White Paper, the UK government recognises that the *Modernising Government* agenda must go forward with the confidence of the citizen, and makes clear its view that 'data protection is an objective of information age government, not an obstacle to it'.[4] This view is backed up by a commitment to work with the Data Protection Registrar to ensure that the privacy implications of electronic service delivery are fully addressed. This welcome proactive privacy stance has yet to be manifested in any telemedical applications under development, as little in the way of advice has been sought from the Data Protection Registrar as yet.

Understanding confidentiality in telemedicine

Whilst those promoting the opportunities afforded by telemedicine are quick to recognise that there are potential security and possibly wider data-protection issues, it is noticeable that little substantive work has been undertaken on patient reaction or what the necessary safeguards should be. Often those citing positive patient reactions to participation in telemedicine appear to do so on the basis of a fairly unsophisticated sounding of views. As with any opinion poll, much depends on the questions that are asked when judging the weight that can be attached to the response. For instance, if a patient is asked for her views about on-line consultation from her GP's surgery with a consultant in a hospital many miles away, her initial reaction may be favourable in that it has saved her a day off work and an arduous journey. However, if it is then pointed out that, unbeknown to her, a roomful of medical students was viewing the consultation she may be slightly more concerned. Similarly, if the consultation was being recorded for potential future use beyond the confines of the consulting room, her concerns may be heightened. To then reveal that the link between the GP's surgery and the hospital is not secure and is readily capable of interception may cause her to change her initial opinion. Thus without a clear exposition of the potential dangers it is difficult for a survey to elicit and reflect the informed views of individuals.

Clearly, one important way to establish and maintain public confidence in telemedical applications is to ensure that data-protection safeguards are identified, put in place and then rigorously complied with. Those who are quick to claim a positive patient reaction are equally quick to assert that developments will, of course, be accompanied by these 'appropriate safeguards'. However, it is a concern that such assurances are often given without any clear thoughts about what the safeguards should be, and those who have addressed the issue often get no further than emphasising the concerns surrounding the security of communications over a public telecommunications network. In order to provide a comprehensive framework of safeguards it is necessary to address all aspects of data-protection legislation. Some of the salient features are described below in the context of the requirements of the Data Protection Act 1998.

Data Protection Principles

At the heart of all data-protection legislation are the legally enforceable standards to which data controllers must adhere. In the UK's Data Protection Act 1998 these are the eight Data Protection Principles. The Principles are set out in Schedule 1 of the Act. Briefly, they require that:

1 Personal data shall be processed fairly and lawfully and in accordance with Schedules 2 and 3
2 Personal data shall be obtained for specified and lawful purposes and not further processed in an incompatible manner

3 Personal data shall be adequate, relevant and not excessive in relation to those purposes
4 Personal data shall be accurate and, where necessary, kept up to date
5 Personal data shall not be kept longer than is necessary for the purposes for which they are processed
6 Personal data shall be processed in accordance with the rights of data subjects
7 Appropriate security measures shall be taken in respect of personal data
8 Personal data shall not be transferred outside the European Economic Area (EEA) unless there is adequate protection for the data in the receiving state.

At first sight many view these principles as a statement of good information-handling practice, and indeed they are, but to describe them in such a way risks underselling the rigour with which they should be complied with. These are not merely for the virtuous, but are legal requirements of all data controllers backed up by an enforcement regime under the control of the Data Protection Commissioner.

The principles in practice in telemedicine
Fair processing

In the UK these principles have been strengthened compared to those under the previous 1984 Act. For example, the first principle still requires that personal data are processed fairly and lawfully, but there are important changes. The requirement that the processing is in accordance with general rules of law (such as the common law duty of confidence) remains, but the principle now expressly provides that personal data are not to be treated as processed fairly unless, as far as is practicable, certain criteria are met. These include informing the data subject of the identity of the data controller and any nominated representative, as well as informing the data subject of the purposes for which the data are to be processed.

It is therefore necessary to ensure that the individual is fully aware at the time of a telemedical interaction of the identity of the person who is processing the data, what it is to be used for, and whether any wider disclosures are envisaged. If a telephone consultation is tape-recorded or the calling number is retained, are individuals informed of this before they start to provide personal details? In order to address not only these fair processing requirements but also those of lawfulness arising from the duty of medical confidentiality, it will be necessary for transactions to take place on the basis of fully informed consent. The Act's requirement to process fairly also means that the actual processing itself must not result in unfairness. For example, in a remote-diagnosis application, a decision-support system may be used to assist in the collection of relevant information and suggest courses of action to an operator. If such a system makes incorrect suggestions because of flaws in the software, and there is a detrimental impact on an individual as a result, this could amount to unfair processing.

One new feature is the Act's requirement that there is a legitimate basis for processing (contained in Schedules 2 and 3 of the Act). Without compliance with this precursor, no processing can take place. If medical data, which are deemed to be sensitive under the Act are processed, the appropriate basis would be that the processing is necessary for medical purposes and is undertaken by a health professional or someone who owes the same duty of confidentiality as a health professional. Establishing a proper basis for processing of all items of data is an essential consideration before telemedical processing commences.

Lawful purposes

The second principle requires that data are obtained only for specified and lawful purposes and are not used in any way that is incompatible with those purposes. Wider use of patient information (for example, to demonstrate the benefits of telemedicine applications at a national conference, as has been witnessed by members of the Registrar's own staff) may be somewhat incompatible with the original purpose of processing in addition to the other concerns about fair and lawful processing which such an activity would raise.

Adequate but not excessive

The powerful potential for computer technologies to store large quantities of information means that it is often difficult for those utilising them to resist the temptation to make often ill-considered decisions to collect and retain information merely because the capacity is available. Compliance with the third principle requires greater self-control, as only data that are strictly necessary and relevant to the purpose should be processed. The principle also requires that it is adequate for its purpose. Is the information processed in a telemedical transaction that is up to the job in hand? If data are incomplete or there are interfaces with other existing systems with differing data-quality standards, does this affect the service that is being delivered? If diagnoses are reached on the basis of incomplete information, or if records of dubious quality or inconsistent data standards have been matched together, then compliance with the third principle would be difficult. Misattribution of data can have even more serious practical consequences for the patient.

Accuracy

It almost goes without saying that accuracy of data is essential. However, what systems are put in place to ensure accuracy? These can be simple measures such as automatic verification of certain features, such that a plausible date of birth has been given, but in some cases whole systems of work may be necessary to underpin accuracy. For example, a consultation over the telephone in which an individual purports to be someone else could have serious consequences if data are added to another person's record. Verification checks may be necessary, and this may involve the need to agree personal identifiers known only to an individual, as is common in the telephone-banking context.

No longer than necessary

The fifth principle's requirement that data are held for no longer than is necessary also requires discipline at the system-design stage. Retention periods should be set for data based on future necessity over time. Continued retention on a 'may come in useful' basis is not acceptable. Judgements should be made about the actual need in the future. In cases where data are captured at both ends of a telemedical transaction, questions may arise as to whether both parties need to retain all of the information recorded. Systems should be designed to allow for differential deletion of data items to permit 'weeding' of specific items of data which have outlived their usefulness and are not preserved because of any other requirement to retain specific records.

Data subject rights

Individual rights are strengthened under the new data-protection regime. In addition to the obligations placed on data controllers, there are specific rights given to individuals. The one that most frequently comes to mind is the right of access (indeed, on a general note, existing access rights under the Access to Health Records Act 1990 are brought within the scope of the Data Protection Act 1998, including the 1998 Act's enforcement and complaints-handling regimes). Individuals will be able to request data processed about them in telemedical applications, which may result in the need to provide images and other such data. If telemedical data are recorded at a remote location, individuals will need to know how to gain access to them, and arrangements must be put in place to facilitate this. Different judgements over access to data identifying third parties are also needed under the new legislation, requiring the balancing of competing interests rather than a 'blanket withholding' of third-party details in records. Ways of dealing with access requests from individuals and for making judgements about the data to be provided will have to be established in the context of each telemedical application.

Individuals' rights do not stop with access. A new right to prevent processing is also granted. This right is limited to processing that would cause substantial unwarranted distress to the individual, and there are exceptions to this right (e.g. where previous consent to the processing has been given), but the possibility of such an action by an individual must be considered when deploying telemedicine applications. In a similar vein, individuals can serve a notice requiring that no decisions significantly affecting them are taken based solely on automated processing. Again there are exceptions to this (e.g. incorporation of adequate safeguards), but all of these matters must be considered should we ever reach the stage where human intervention is not present in the decision-making process.

Individuals also have rights to claim compensation for damage and distress they suffer as a result of any contravention of the 1998 Act. This is a large expansion of existing compensatory rights, and individuals can also ask a court to rectify, block or erase data processed in contravention of the Act. These are in addition to the enforcement regime at the disposal of the Data Protection Commissioner.

Appropriate security measures

Security is often the first and only data-protection issue to be considered in the debate on telemedical developments, and it is indeed an important feature. Security standards have been strengthened in the Act's seventh principle. When judging the appropriate level of security to deploy, it is no longer sufficient simply to consider the nature of the data and the harm that may be suffered by an individual as a result of any breach of security. The 1998 Act now requires that the state of technological development and cost are taken into account. Techniques that were costly only a matter of months ago are now becoming increasingly affordable. Given the sensitivity of health data, it is difficult to see how telemedical applications can proceed without a high standard of security. Some of those demonstrating telemedical applications are keen to show that the cost of equipment may be low enough to allow for on-line consultations with a housebound patient, possibly using dated and increasingly obsolete second-hand equipment. Care needs to be taken to ensure that the equipment is adequate to support the necessary levels of security.

There is a danger that security may be viewed solely as a series of technical issues such as encryption of data. However, the whole system of work needs to be considered. Introducing technology into a new environment can pose a host of problems ranging from the information-handling culture of those using it through to the physical surroundings in which it is to be located. Even those developments using telephone call-centre approaches need to be mindful about treating security as an issue in the context of the general system of work. If patient records are to be accessed and consulted, what level of verification is in place to ensure that the person on the other end of the telephone is the subject of the record? Is there sufficient information in the record which would be known only to the patient and which could be utilised to verify the identity of the caller? If not, it may be necessary to introduce specific identity-verification controls, such as allocating passwords as is done in telephone banking.

Restrictions on transfer

One further significant area of change in the legislative approach reflects the concern that standards for processing personal data elsewhere in the world are lower than those set within the European Union. The eighth principle states that personal data may only be transferred outside the European Economic Area if an appropriate level of protection exists. There are exceptions to this provision (e.g. cases where transfers are made with consent), and much work is being done at present to resolve potential difficulties that may arise with some of the member states' major trading partners (e.g. the USA) due to the lack of legislative safeguards in such third countries. Those developing a telemedical application in which data are transferred to, say, a consultant at a US hospital may, for example, wish to obtain specific consent to the transfer or else consider whether there is adequate protection in the recipient country.

Safeguards, not stumbling blocks

Initial consideration of the previous points may provoke various responses. One opinion may be that the data-protection safeguards are onerous and difficult to deploy in practice. Another response may be to question whether most of the issues are not just matters of common sense. It is wrong to assume that there is anything unique to tele-medical applications which causes insurmountable data-protection difficulties. The data-protection issues facing telemedicine are not fundamentally different to those that have already faced other activities using similar techniques, and which have been successfully addressed in other such areas. Many of the data-protection considerations are a matter of common sense – legally enforceable common sense. No one would wish to deploy a system that will hold inaccurate data or excessive data, but delivery of this common-sense aim requires fore-thought and cannot be left to chance alone.

New technological applications bring with them not only opportun-ities for achieving certain operational goals, but also the chance to deploy technological solutions to protect personal privacy. These are increasingly known as privacy-enhancing technologies. The technolo-gical possibilities for reducing the amount of patient-identifiable data in transactions accord not only with the Data Protection Act but also with the recommendations contained in the Caldicott Report,[5] which reviewed the use of patient-identifiable data within the NHS. However, opportunities have to be seized. The opportunities afforded by tele-medicine to improve levels of service to patients and to maximise the use of hard-pressed resources are now being identified and are starting to be seized, but the accompanying possibilities for doing this in a way that maximises personal privacy are not being grasped with the same vigour. The confidence of patients in their health professionals and in the healthcare system is currently high, although recent and as yet unpublished research by the Data Protection Registrar into public attitudes to privacy and data protection within the UK shows that this high level of confidence is declining in the healthcare sector in common with other areas of the public sector.

No one would want a situation to arise in the future where patients look back to paper medical records and the face-to-face consultation in the home, hospital or doctor's surgery as a golden age of personal privacy. Developments in telemedicine must only go forward, taking into account the patient's right to privacy and the need to comply with the legal enforcement regime which underpins these rights. Those developing telemedical applications should reflect on the current UK *Modernising Government* agenda and adopt a similar stance, recognising that data protection is an objective of telemedicine, not an obstacle to it. The opportunities to protect personal privacy are there to be grasped. An individual's personal privacy and confidence in those who provide them with healthcare are matters which share a potential but unwelcome possibility – that once lost, both are difficult if not impossible to restore.

References

1 Wootton R (1995) *The benefits of telemedicine.* In: Conference abstracts: Exchanging Healthcare Information. BJHC Ltd, Weybridge, 69–70.
2 Grigsby G and Sanders JH (1998) Telemedicine: where it is and where it is going. *Ann Intern Med.* **129**(2): 123–7.
3 Cabinet Office (1999) *Modernising Government.* White Paper. Cmd 4310. The Stationery Office, London.
4 Cabinet Office (1999) *Modernising Government.* White Paper. Cmd 4310, Ch. 5, para 14. The Stationery Office, London.
5 Department of Health (1997) *The Caldicott Committee. Report on the Review of Patient Identifiable Information.* Department of Health, London.

CHAPTER 12

Commissioning health telematics: the issues to be addressed

Judith Greenacre

Introduction

There is an increasing role for health telematics applications across the spectrum of healthcare, including not only direct patient care but also education, research, administration and public health. Some applications are already accepted without question in daily practice, whilst others continue to capture the imagination of the public and attract considerable media interest. The extent and speed of development of specific telehealth applications vary considerably between and within countries, as commissioners and providers of healthcare adopt varying perspectives on the scope offered by health telematic applications and on the potential costs and benefits arising from their use. Commissioners of health services have a key (but not unique) role to play in ensuring that health telematic technologies are appropriately incorporated into service provision. The benefits and opportunities presented by health telematics are described elsewhere. This chapter focuses on the issues that the emerging technologies present to health service commissioners in the National Health Service (NHS) in the UK.

The scope of health telematics

Much has been written about definitions and terminologies associated with health telematics. Most definitions embody the concepts described in the World Health Organization (WHO) definition.

> *'Health telematics is a composite term for health-related activities, services and systems, carried out over a distance by means of information and communications technologies, for the purposes of*

global health promotion, disease control and health care, as well as education, management and research for health.'[1]

The WHO further defines different activities under the health telematics umbrella to include telemedicine, tele-education for health, telematics for health research and telematics for health services management. However, for all of these the basic concepts described in the WHO definition remain the same.

Although it is useful to be able to describe health telematics succinctly, these definitions do not immediately convey the scope and potential of the technologies to commissioners of healthcare. In a recent study in West Wales,[2] service-orientated examples of the potential scope of applications within a local NHS Trust were identified, as shown in Figure 12.1. Such practical illustrations of the uses of health telematics can be more helpful to commissioners than the standard definitions.

Direct patient care			Clinical support		
Health monitoring	Remote consultation	Care planning	Patient information	Education	Management and administration
Virtual home visit Telephone triage Patient condition monitoring	Interactive video consultation Electronic store and forward consultation Voice messaging	Automated workload scheduling Multi-disciplinary case co-ordination Multi-agency care co-ordination On-line access to pathways Knowledge-base access to experts	Community electronic patient record Medical information support service	Voice conferencing Video-conferencing education Computer-based training for clinicians Computer-based training for patients Freephone health information service	Programme evaluation Service utilisation monitoring Contact logging Care pattern analysis

Source: Ceredigion and Mid Wales NHS Trust

Figure 12.1: Potential telematics applications identified in the 'Keeping Care Local' Study, April 1998.

Who are 'the commissioners' for health telematics?

The verb 'commission' can be defined *as 'to give a commission to or for; to empower; to appoint; to put into service'*.[3] Thus with regard to health telematics, health service commissioners could include anyone who has some responsibility for putting the technology into service.

In contemporary Britain, 'health service commissioners' have recently been thought of as health authorities and general practitioner (GP) fund-holders. The roles of commissioners are changing as fund-holding ceases and as primary care groups take on commissioning responsibilities both from them and from health authorities. In addition, a stronger role is emerging for local government to shape health services both through their involvement in primary care groups and through strengthened public health responsibilities.

NHS Trusts (the traditional providers of services) have their own role in planning and developing their services in response to their commissioners' strategic plans. This role has been considerable, especially in areas where there has not been one main commissioner and/or where they have had the monopoly on service provision. Among Trust staff are most of the prime enthusiasts for and clinical advocates of health telematics. Thus Trusts will continue to play a key role in determining and developing the use of health telematics in everyday practice.

National and regional health service managers have a key commissioning role through setting the strategic direction for health services in response to current political thinking, and through implementing performance management in the service.

Public expectations and concerns can influence commissioners at every level. Public acceptance of the technologies is central to the commissioning process.

The need for wider co-ordination

By their very nature, health telematics applications may impact contemporaneously on the work of several commissioners. If developments are to be implemented successfully, they therefore need to be planned and considered by an appropriate range of commissioners, rather than by one or two in isolation from the rest. Indeed, failure to take on board the range of agencies and the levels within them from which support is required may have been a key factor in previous failures to implement health telematics developments successfully.

Ensuring that health telematics is fully utilised will require not just the motivation of traditional 'commissioners', but also broad culture changes at many levels. Such change is unlikely to occur without proper discussion of the issues that health telematics poses and credible evidence of the benefits to be gained.

Commissioning issues

Strategic and planning issues

The changes outlined in the current NHS reforms, together with the financial challenges already facing the NHS, mean that the current commissioning agenda is vast. Urgent and important issues that are currently uppermost in the minds of commissioners include the following:

- the development of primary care groups as future commissioning bodies
- the abolition of the internal market and changes to contractual frameworks
- developing health improvement programmes, whilst
- attempting to manage additional resource challenges posed by developing technologies, the new public health agenda, etc.
- the implications of clinical governance
- the implications of devolved government in Wales, Scotland and Northern Ireland
- continuing mergers of NHS Trusts and of health authorities
- continuing requirements to meet performance management targets
- collaboration and partnership between agencies within and beyond the NHS
- addressing the poor quality of information in the NHS, coping with Year 2000 challenges, etc.

In the face of such an agenda, commissioner views on health telematics range from it being an expensive and unproven additional development, through to it providing imaginative innovative solutions to health service management problems, as well as improving patient care.

Scarcity of evidence

There is a lack of good-quality information about the relative costs and benefits of using health telematics for health service delivery (*see* below and Chapters 2 and 5). Many commissioners may not have thought about the potential for telehealth to help deliver their health service commissioning agenda.

In the USA, competing priorities and the lack of robust evidence of relative benefit, together with poor infrastructure, planning and development mechanisms have all been identified as key barriers to the introduction and sustainable use of telemedicine.[4] Some research has recently been undertaken to establish good practice for planning and implementing telemedicine developments,[5] but further work is needed to develop and disseminate these ideas.

Politics, priorities and partnerships

Political interest in health telematics is increasing, as is evidenced by its high profile in national information management strategies.[6,7] However, flexibility in delivering other imperatives may be required if time and resources are to be diverted from other commissioner priorities and service developments in order to develop appropriate telematics commissioning.

Resource issues for health telematics need to be addressed seriously and imaginatively. Partnerships with other public service partners and the private sector should at least be considered, together with more innovative financial initiatives (e.g. 'spend-to-save' schemes in which initial additional funding is made available to set up schemes that are expected to deliver savings in the medium to longer term).

Medico-legal issues

Various medico-legal issues relating to health telematics (particularly telemedicine) have been identified, and these are described in detail both in the preceding chapters and elsewhere.[8,9] Some of the key issues of relevance to commissioners are summarised below. Commissioners can help with some of these by ensuring that purchased services are governed by appropriate agreed protocols that seek to avoid some of the pitfalls. Examples of good practice should be widely disseminated and might include, by way of example:

- ensuring that the individual with overall clinical responsibility for the patient is identified at the outset of a teleconsultation procedure
- ensuring that, for teleconsultations involving experts outside the UK, the clinician is accredited with specialist training and experience comparable to that expected of specialists practising in this

country. Experts from overseas should also be adequately insured against litigation as a result of the consultation.

Litigation

Widespread concerns have been raised in the USA over potential litigation arising from technical failures or diagnostic errors that result from a teleconsultation rather than 'face-to-face' consultations. It has been suggested that the risk of litigation is lower in cases where the practitioner maintains an advisory role rather than a doctor–patient relationship.[10,11] The American College of Cardiology notes that litigation is less likely if a good doctor–patient relationship exists.[8] This relationship is certainly altered by telemedicine consultations although the effect on litigation remains unclear. To early 2000, there have been no reported successful malpractice claims in the USA associated with telemedicine, although the number of teleconsultations that have taken place remains small. As a product of the technology, consultations are often recorded, raising the potential to assess liability, albeit with the benefit of hindsight. This could either reassure clinicians or deter them from developing telemedicine practice in the future.

Accreditation

State-based licensure to practise medicine in the USA has severely limited the interstate practice of telemedicine,[8] despite the fact that examination of physician competence and developments of practice standard guidelines occur nationally rather than state-wide. Some of these difficulties have also been described in Europe since the signing of the Maastricht agreement.[12] Academic institutions, governments and intergovernmental bodies should consider the development of national and international accreditation procedures for telemedicine consultations.

Security and confidentiality

General issues relating to the confidentiality and security of patient-identifiable information are discussed in detail in Chapter 11 and elsewhere.[13] These are not unique to health telematics, and opinions vary as to whether information in electronic format offers more or less security than paper-based information. Much has been done to improve the security and confidentiality of patient-identifiable data (for example, introduction of the Data Protection Acts, European Directives on patient confidentiality, the adoption of the Caldicott principles,[14] etc). However, both general issues and those specific to health telematics will need further discussion before all current concerns are resolved.

Consent

Patients need to be properly informed about the ways in which health telematics may impact upon their treatment and/or their relationship with their clinician(s). The extent to which information about themselves may be recorded, edited and shared between professionals also needs to be explained, no matter what safeguards are in place to keep it secure and confidential. Guidelines have been produced for informed patient consent to the use of identifiable data.[15] The potential to share information is bound to rise with increased implementation of health telematics. This can bring benefits both in terms of improved access to information and in terms of data quality. However, the implications need to be understood and agreed upon both by individuals and by the public as a whole. Further work in this area is undoubtedly required.[16]

Research and ethics

Telemedicine consultations can change the relationship between patient and clinician, although further research is required to explore and assess these changes.[11] The ethical implications of such changes will need to be debated as work in this area progresses.

Health authorities have a statutory duty to ensure, through their Local Research Ethics Committees (LRECs), that health research involving their resident populations is ethically acceptable. Health telematics research is no exception to this requirement. Telematics brings a new range of scenarios for LRECs to consider, and members may require additional guidance on the ethical issues which may arise from such research.

The requirement to approach individual locally based ethical committees can present major difficulties to multicentre, multidistrict or multinational research projects. This is likely to be a particular challenge for health telematics research, which by its very nature often crosses organisational and national boundaries.

There may be a case to be made for larger specialist ethics committees to be set up to consider health telematics research, thus relieving some of the difficulties for both LRECs and researchers.

Financial issues

The problems of limited financial resources to be allocated to many competing health service priorities have already been highlighted. As mentioned below and in earlier chapters, along with other aspects of evaluation of health telematics, there is a lack of robust economic analyses of costs and benefits.

The absence of agreed fee structures for telemedicine consultations has presented a significant barrier to telemedicine developments in the USA.[4,5,8,11,17] Similarly, the coding systems to ensure that activity is recorded in a uniform way are inadequate. NHS funding arrangements make these issues less of a problem in the UK, but problems still remain in ensuring that telehealth activities are uniformly recorded and costed.

The costs of health telematics have fallen considerably over the past decade, but they continue to represent a substantial resource. Initial costs are high, and potential future savings are often not accrued by those who met the initial costs. Financial factors may well influence the course of future telemedicine development. For example, many telemedicine providers are moving away from expensive interactive video telemedicine systems and towards cheaper desktop 'store-and-forward' systems.[17]

Evaluation issues

The many cited benefits of health telematics include improved access to care, reduced travelling costs, enhanced clinical knowledge and skills, patient acceptability and networking between clinicians. However, there remains a paucity of comprehensive evaluations of the costs and benefits of health telematic applications. Also lacking are comprehensive analyses of the infrastructure implications and financial requirements for sustaining telemedicine.

Methodologies

Where evaluations have been undertaken, systematic approaches to studies have often been lacking. The following are examples of practice that commissioners would find helpful.

- The questions to be answered should be clearly defined and agreed upon.
- Distinctions should be drawn between assessments of technical

feasibility, benefits to patients, benefits to clinicians and improved quality of service.

- Evaluations of resource costs and benefits must be robust and include opportunity costs as well as an indication of flows of costs and benefits between key players.
- Assessments of technological issues should be kept as straight-forward as possible. Complex technical issues could be set out in a separate technical document.
- Attempts should be made to distinguish between potential short-, medium- and long-term impacts of the application(s) under study.
- Alternative explanations should be sought for observed changes. For example, a reduction in a GP's dermatology waiting-list should not be ascribed to the previous introduction of a teledermatology consultation service without considering the trend in waiting-lists elsewhere, the impact of waiting-lists initiatives, etc.

Progress is being made in developing appropriate tools for assessing telecommunications in healthcare.[18–20] Such examples of good practice should be widely disseminated and their use incorporated into future evaluations of health telematics. Only in this way will commissioners, providers and the public eventually obtain the information that they require to inform their decision making.

Outcome measures

Health telematics embodies new methods of service delivery rather than new treatments. Outcome measures for evaluating telematics need to take into account the method of service delivery rather than the effects of the treatments or programmes themselves, and they need to be able to measure against the non-telematic method of delivery that would otherwise be used. Evaluations may be required at the following levels:

- an individual level (e.g. the benefit to a GP of gaining access to an electronic noticeboard)
- an organisational level (e.g. the benefits of introducing electronic messaging throughout a hospital);
- a population level (e.g. to evaluate the impact of patient information delivered via publicly accessible information kiosks).

The development of such measures that can effectively measure the impact of health telematics remains an ongoing challenge for the twenty-first century.

Quality assurance

Having evaluated the effectiveness of health telematics services and incorporated them into everyday practice, there is a need to ensure that the quality of the resulting service is maintained at an appropriately high level. Quality assurance presents many challenges to health services. In the case of health telematics there is a need to tease out assessment of the quality of the telematic system from assessment of the clinical event being delivered.

Models to assess health service quality

There have been many attempts to define and model concepts of quality in relation to health services. Donabedian's analysis of organisational structures, processes and outcomes[21] has been pivotal in the development of quality assurance measures. In recent years there has been an increased focus on outcomes, which are usually considerably more difficult to assess than measures relating to structure or process. Other useful analyses include Maxwell's six dimensions of the quality of a service (effectiveness, efficiency, equity, acceptability, accessibility and relevance)[22] and that provided by the World Health Organization, which divides quality into the following four aspects:

- professional performance (technical quality)
- resource use (efficiency)
- risk management (the risk of injury or illness associated with the service provided)
- the patient's satisfaction with the service provided.[23]

Regardless of the framework, the drive to develop processes in order to improve, assure or manage the quality of health services has been described as essentially involving the following:

- agreeing upon the desired attributes of any given type of service
- establishing ways of working to produce services with these attributes as efficiently as possible.[24]

Audit

The clinical audit cycle that was developed during the 1980s provided a simple quality assurance model that could be used among peers within the NHS. It involves setting standards, monitoring performance and instituting changes so as to improve performance in areas where standards have not been met.

Although peer review is established in the UK, the implementation of clinical audit has not been a complete success. Clinical failures have not been detected, participation in audit has been incomplete and declining, and there has been poor communication with managers.[25]

Clinical governance

The difficulties of assuring the quality of health services have been reviewed in the recent shadow of much-publicised failures to maintain appropriate standards of care and noted unacceptable variations in clinical practice. This has led to a recent strengthening and development of the NHS quality assurance function.[26,27] Health organisations will now have a statutory duty to seek continuous quality improvement through clinical governance, and chief executives will be held accountable for ensuring that this develops at every level within their organisation. New national bodies are being set up to help to define and ensure service quality management. The NHS will be expected to address issues relating to clinical quality with the same vigour that has previously been devoted to service performance and financial management.

Clinical governance has been defined as 'a system through which NHS organisations are accountable for continuously improving the quality of their services and safeguarding high standards of care by creating an environment in which excellence in clinical care will flourish.'[25] The process will continue to rely on local professional self-regulation to address poor clinical performance, and will promote the recognition and replication of good clinical practice with lessons learned from failures as well as from successes. To promote this way of working it will be particularly important to develop strong clinical leadership and positive organisational cultures and values.

Health telematics and quality assurance

Clinical governance means that assuring the quality of services is now a necessity rather than an option. As new health telematics applications emerge, quality measures will need to be defined and demonstrated. Rather than being perceived as a drawback, this should be seen as a golden opportunity to establish the value of telematics in modern health service delivery.

Unfortunately, to date there has been relatively little work on developing quality assurance programmes in health telematics. Although the number of Internet websites relating to health telematics is legion, less than 1% are selected on searches that include the term 'quality assurance'. A similar picture is also seen for published articles (personal observation, August 1999).

Where research has been undertaken, it has concentrated on measures of process, acceptability and patient satisfaction. Further work is required, especially with regard to assessment of outcomes, effectiveness, resource use and risk management.

Relatively simple matrices can be developed to assess the quality of health telematics services, using those frameworks already described for other aspects of health service quality assurance.[28,29] The quality of the clinical and technical aspects of a health telematics service should be assessed in a manner similar to that used for non-telematic clinical events. Additional safeguards may be required to ensure that the usual standards are not being compromised by delivery through a telematics system, as illustrated by the following extract.

> 'The teleradiology image at this facility is poor in at least 50% of the cases presented to me. In many instances, the image quality is not much better than Dr Roentgen's original radiographs of almost 100 years ago. It is virtually impossible to have images reported or transmitted . . . by the time we receive them, the technologist . . . is long gone from the facility where the exam was done.'[30]

Work has been undertaken to develop quality-control processes with regard to health information that is available on the Internet.[31,32]

Research on the safety and efficacy of telemedicine systems has been reviewed and found to fall into the following three phases.[33]

1 Identify the technical specification of equipment required for the application.
2 Test that this is appropriate in particular settings.
3 Establish a set of standards and guidelines to ensure that the system is used to best advantage.

Although some technical standards for telematic equipment have been developed, reviews of published works indicated that most research was in phase 1 or 2, and that the quality of research needed to be improved.

In the UK, a culture change is needed to enable the quality of health services to be assured adequately. These changes must occur at every level and will also need to extend to those working to develop health telematics applications. Professional bodies are starting to consider the indices by which the quality of health telematics applications might be judged,[34,35] but further work is needed both nationally and internationally to develop and agree upon appropriate standards and quality measures. The new organisation-wide approach towards developing quality in the NHS provides a welcome umbrella under which this work can be co-ordinated and developed.

Ways forward for commissioning health telematics in the National Health Service

Advances in telecommunication technologies mean that the question of 'whether' health telematics has a role in modern healthcare is redundant, and has been replaced by the questions relating to the speed and extent of implementation. Change is inevitable and a balance must be struck. The National Health Service should not be unnecessarily inefficient through persistent reliance on old ways of doing things, nor should it waste scarce public resources on the widespread implementation of new technologies in the absence of a proper evidence base, exacerbated by inadequate planning and evaluation mechanisms. The way forward is not easy either for commissioners of health services or for those in the health telematics industry, but efforts have been made to chart the way forward.[36] Planning the development of health telematics should include the following.

- *A focus on real needs for health telematics.* The industry needs to listen to what the service is saying and focus developments on needs rather than on high-cost 'one-off' developments that can hit the headlines.
- *Maximum use of applications.* NHS agencies need to ensure that they invest in applications which can meet their assessed priority needs. However, once installed, use of applications should be maximised by consideration of other potential uses and, most importantly, adequate ongoing training and support of staff.
- *Thorough evaluations of health telematics applications.* The body of evidence of the effectiveness of many health telematics applications needs to be expanded. Evaluations must include consideration of all the costs and benefits in an intelligible format. This should include honest appraisal of the failures as well as the successes, as all involved can learn from others' mistakes. Inadequate evaluations do nothing to increase confidence in the technology among commissioners, and may indeed delay wider implementation.
- *Planning for change.* Commissioning is as much about change management as it is about the technologies themselves, and will not succeed unless the following criteria are met.
  1 *All key players are appropriately involved.* Multi-agency planning includes all who have a role in implementing the service. Such

planning will also assist uniformity of technical specifications
and systems compatibility

2 *Environments for change are created.* To facilitate the changes,
flexibility may be required in the current commissioner performance priorities. Many commissioners are challenged to develop
services in the face of mounting resource constraints. If the
political will to develop health telematics exists, more imaginative ways to fund and support developments are required than
simply asking commissioners to prioritise them against other
elements of local service provision.

3 *Real and relevant benefits are meaningfully demonstrated.* Individuals need to be able to see for themselves what the technologies can offer to them. The examples shown in Figure 12.1
illustrate the many potential applications concerned with patient
support rather than direct patient care. Such support applications
may not appear to be as exciting as some of the more unusual
direct care applications, but they may well represent more tried
and tested technologies that could be implemented relatively
easily and that could demonstrate the real opportunities for and
benefits of telematics in the future NHS.

4 *The quality of health telematic services is properly defined and
assured* from the outset.

Conclusion

Health telematics offers many innovative ways of improving the
quality of healthcare, both through direct clinical services and through
indirect support such as medical education and management. Commissioning affects all of those who are involved in putting the technology
into service, and for this to be successful, changes in traditional
cultures and working patterns are needed. As part of these changes,
more robust evidence of the effectiveness and efficiency of applications
is required. For such evidence to be collected, evaluation methodologies
for health telematics need to be refined and expanded.

Commissioners also need to consider the wider medico-legal issues,
including accreditation of experts across international boundaries,
security and confidentiality of data, fully informed patient consent,
potential litigation scenarios and ethical considerations. Consistent
with other changes in the NHS, the quality of health telematics service
delivery needs to be properly defined and assured.

Much work needs to be done in many of these areas, and this may
seem daunting in the face of the many other competing health service

pressures and priorities. However, there can be no doubt that health telematics holds the key to many exciting new routes for health service delivery in the twenty-first century and beyond.

The views expressed in this chapter are the author's, and do not necessarily reflect those of her employing authority or of any other agency mentioned.

References

1 World Health Organization (1997) *A Health Telematics Policy in Support of WHO's Health-For-All Strategy for Global Health Development*, 11–17 December 1997, Geneva.
2 Ceredigion and Mid Wales NHS Trust (1998) *Keeping Care Local Feasibility Study. Executive Summary and Summary Report*. NHS Wales, Cardiff.
3 Larousse plc (1994) *Chambers Dictionary*. Larousse plc, Edinburgh.
4 Souby JM (1995) *The Western Governors' Association Telemedicine Action Report*. www.arentfox.com/western.htm#vision
5 Joint Working Group on Telemedicine (1997) Evaluation. In: *Telemedicine Report to Congress*. http://www.ntia.doc.gov/reports/telemed/evaluate.htm
6 Burns F (1998) *Information for Health: an information strategy for the modern NHS 1998–2005*. NHS Executive, Leeds.
7 Welsh Office (1999) *Better Information: Better Health. Information management and technology for healthcare and health improvement in Wales 1998–2005*. NHS Wales, Cardiff.
8 *Ad-hoc* Task Force on Telemedicine White Paper (1997) *The Cardiovascular Specialist and Telemedicine*. American College of Cardiology, Bethesda, Maryland.
9 Stanberry B (1998) *The Legal and Ethical Aspects of Telemedicine*. Royal Society of Medicine Press, London.
10 Egan M (1997) Trust in telemedicine: online ethics, security and data. *Med Imaging Monitoring*. **May**: 24–5.
11 John Mitchell & Associates (1998) Telemedicine policy issues (Topic number 2). *JMA Discussion Paper on Telemedicine*. http://www.jma.com.au/telemeddp2.htm
12 Shannon GW (1997) The Atlantic Rim Telemedicine Summit. *Telemed J*. **3**: 269–96.
13 Darley B, Griew A, McLoughlin K *et al.* (1994) *How to Keep a Clinical Confidence: a summary of the law and guidance on maintaining the patient's privacy*. HMSO, London.
14 The Caldicott Committee (1997) *Report on the Review of Patient-Identifiable Information*. Department of Health, London.
15 Welsh Office (1996) *Guidance on the Protection and Use of Patient Information*. NHS Wales, Cardiff.
16 Rigby M (1999) The management and policy challenges of the globalisation effect of informatics and telemedicine. *Health Policy*. **46**: 97–103.
17 Grigsby J and Sanders J (1998) Telemedicine: where it is and where it's going. *Ann Intern Med*. **129**: 123–7.
18 Puskin D, Brink L, Mintzer MPA *et al.* (1995) Joint federal initiative for creating a telemedicine evaluation framework. *Telemed J*. **1**: 393–7.

19 Field MJ (1996) *Telemedicine: a guide to assessing telecommunications in healthcare.* National Academy Press, Washington DC.

20 Friedman CP and Wyatt J (1997) *Evaluation Methods in Medical Informatics.* Springer-Verlag, New York.

21 Donabedian A (1980) *Explorations in Quality Assessment and Monitoring. Vol. 1. The Definition of Quality and Approaches to its Assessment.* Health Administration Press, Ann Arbor, MI.

22 Maxwell R (1984) Quality assessment in health. *BMJ.* **288**: 1470–2.

23 World Health Organization (1983) *The Principles of Quality Assurance.* World Health Organization, Copenhagen.

24 Taylor D (1996) Quality and professionalism in health care: a review of current initiatives in the NHS. *BMJ.* **312**: 626–9.

25 Scally G and Donaldson L (1998) Clinical governance and the drive for quality improvement in the new NHS in England. *BMJ.* **317**: 61–5.

26 Secretary of State for Health (1998) *A First Class Service: quality in the NHS.* The Stationery Office, London.

27 NHS Wales (1998) *Quality Care and Clinical Excellence.* Welsh Office, Cardiff.

28 Bashshur RL (1995) Telemedicine effects: cost quality and access. *J Med Systems.* **19**: 81–91.

29 Grigsby J, Barton PL, Kaehny MM *et al.* (1994) Telemedicine policy: quality assurance, utilisation review and coverage. In: *Analysis of Expansion of Access to Care through Use of Telemedicine and Mobile Health Services. Report 3.* Centre for Health Policy Research, Denver, CO.

30 Bagby RJ (1996) The need for preservation of quality in telemedicine: state licensure. *J Florida Med Assoc.* **83**: 601–2.

31 Eysenbach G and Diepgen TL. (1998) Towards quality management of medical information on the Internet: evaluation, labelling and filtering of information. *BMJ.* **317**: 1496–502.

32 Wyatt JC (1997) Commentary: measuring quality and impact of the World Wide Web. *BMJ.* **314**: 1879.

33 Taylor P (1998) A survey of research in telemedicine. 1. Telemedicine systems. *J Telemed Telecare.* **4**: 1–17.

34 American College of Cardiology (1999) Quality of care and health guidelines. In: *ACC Health Policy. Section VI.* American College of Cardiology. Bethesda, Maryland.

35 Partnership for Health Informatics/Telematics (1999) *Informatics/Technical Standards.* Canadian Institute for Health Information www.cihi.ca/partship/canhltmd.htm

36 Yellowlees P (1997) Successful development of telemedicine systems – seven core principles. *J Telemed Telecare.* **3**: 215–22.

CHAPTER 13

Effective commissioning of telematics: lessons from the introduction of an integrated clinical communications system in Teesside

Rosemary Taylor

Setting the scene

Teesside is a geographically compact setting with a population of 560 000 inhabitants, and has some of the worst health indicators in the UK. It consists of a complex operational environment with the following:

- one health authority
- four unitary local authorities
- four primary care groups
- 85 general practitioner group practices
- three NHS Trusts
- 90 pharmacists
- 167 dentists
- 124 opticians.

The overall aim of the Cleveland Wide Area Network (CWAN) Project is to enhance patient care and reduce bureaucracy by the introduction of a secure electronic communication system linking GPs, hospitals, community service providers and the health authority. The goal is to establish a Wide Area Network (WAN) within the geographical area of Cleveland using the existing NHS network (NHSnet) as the 'backbone' of the system which will support the flow and exchange of clinical, demographic and administrative data. This will provide support to five major Trusts serving the Cleveland population of 560 000 inhabitants, the 85 general practitioner practices serving them, the emerging primary care groups and the health authority. It does not include Social Services, as the latter are currently outside the NHSnet project.

The project is concentrating on the development of the following:

- e-mail
- structured discharge letters
- edifact pathology results
- non-urgent transport booking.

The challenge, as to all commissioners, was to facilitate an agreed vision and create the environment for change. There was a need to overcome the ingrained attitude of competition by harnessing the energy and commitment of the full healthcare community in creating a whole systems approach to electronic communication (telematics in its widest application).

A learning experience

The explicit vision shared within the NHS natural community, initially driven from primary care, was to enable electronic communications between GPs, hospitals and the health authority, a simple statement that encompassed myriad different themes and tensions. These were:

- interorganisational
- interprofessional
- technical
- behavioural.

The culture of the system had to change from one in which all organisations, clinicians or managers operated independently to one in which the benefits of interdependency were recognised and practised. Individuals and organisations were encouraged to adopt a 'can do' philosophy and to move from being risk averse to being risk aware.

In the environment of the NHS, investment in technology is seen as competing for monies otherwise earmarked for direct patient care. This is in contrast to other industries where technology is embraced as a route to greater productivity. The highly publicised 'failures' of information technology (IT) investment have left a legacy of doubt in the efficacy of IT investment, and the required balance of proof in terms of tangible benefits and value for money is significantly higher than for clinical procedures. The 'blame culture' operating within the NHS has a tendency to encourage risk-averse behaviour.[1] In the area of policy, particularly in relation to national procurement strategy and managing the supplier market, development has lagged behind the needs of the service.

The project

The Tees experience centred on preventing further fragmentation of the IT infrastructure by looking for ways to integrate a plethora of systems that would enable communications via the following:

- e-mail
- structured form-based discharge letters
- edifact pathology results
- ambulance booking service
- Web-browsing.

All of these would occur from the desktop of the general practitioners. This latter point was considered essential if the expected benefits and use of the overall system were to materialise. An enhanced fax machine was not considered to be appropriate. The NHS network (X400) was identified as an agent of change – in each subproject, technology was not the driver but an enabler of a quicker and more streamlined process that would ultimately benefit patients. To achieve the required functionality, it was imperative that the system which was developed was open, flexible and holistic. The platforms adopted should be mature and stable and international standards adhered to. The project challenged some of the previously held views of appropriate levels of functionality which had been based on one-to-one, as opposed to one-to-many, communications. Close working was established between the NHS Executive and other national pilot sites, particularly Avon.

Constraints encountered

As the project progressed, various constraints were encountered. These included policy development in relation to the following.

- *Security and confidentiality.* This was acknowledged as a significant area in which clarity of central policy was underdeveloped. Tees undertook two national security and encryption pilots in which the availability, integrity and confidentiality of the X400 was challenged. In addition, risk assessment workshops focused the attention on the organisationally based risk that contributed to the NHS security manual[2] and training for Caldicott guardians.[3]

- *Pace of change in technology.* Throughout the period of the project (October 1996 to early 1999) the argument has raged over possible selection of the Internet vs. X400 as the technical solution. This is an example of the perennial dilemma of whether to delay in the hope of tomorrow's promise or proceed today on proven technology which could soon be outmoded. In addition, Windows NT has become a *de facto* standard, and clinical systems suppliers have been slow to respond to this.

- *Variety of disparate systems.* Unless commissioners have practised a narrow reimbursement of GP clinical suppliers, an individual health authority is unlikely to have sufficient influence to manage the market effectively. Market share and compliance to standard can only be managed at a national level. A move towards a common operating platform with multiple applications would seem to be a sensible way forward.

- *Procurement.* The scope and appropriateness of framework contracts were considered. When aggregating and co-ordinating the costs of IT across a natural community, the previous capital investment limit of £1 million was encountered, thus negating the benefit of framework procurement. This situation has now been resolved.

- *Value for money (VFM).* As previously outlined, the balance of proof required from VFM is high. However, it is difficult to argue that benefits derived from technology result solely from that investment. Furthermore, benefits are more qualitative than quantitative and do not always align themselves for the auditor's eye. In nationally prescribed targets, benefits should be explicitly outlined nationally in order to negate the need for each separate commissioner to prepare a VFM argument.

- *Professional endorsement.* Reconciling the different perspectives of clinicians in general practice and in hospitals was challenging. Whatever solution is adopted, the selection process needs to have involved clinicians from the outset in order to ensure maximum usage and thus changed behaviour and value for money. Very different perspectives, particularly with regard to security issues, were forthcoming and needed to be synthesised. This local perspective has been mirrored nationally.

- *Standards.* Although international standards were available, they had not been incorporated into hospital system supplier specifications or into the requirements for accreditation (RFA) for general practitioner systems. Edifact standards existed for pathology, but those for radiology and discharge were still in development. Suppliers were keen to offer 'plug-and-play' solutions, but reality did not match their rhetoric.

- *Resource allocation.* With the benefit of hindsight the project was underfunded in both capital and human resource terms. It was not a straightforward implementation but rather a development adventure. The technical skills and expertise available within the natural community were stretched, and the estimated value of investment at £2 million capital was challenging in a cash-limited environment.

Managing change

The dilemma in managing change of this nature is whether it should be approached on an evolutionary basis or by revolution, and there are many schools of thought on this. From a practical perspective it is often a balance. However, modernising of the NHS will require courage and commitment. It will need good research and development programmes that provide the service with costed workable options that both maintain the corporacy of the NHS and allow for local flexibility. Linkages to other public-sector agencies should not be forgotten, and an alignment of policy with regard to health and local authority services is essential.

The expectations of professionals, patients and the public through the boom of home computing, the Internet and the ability to bank, book theatre and travel and gain access to the most up-to-date knowledge remotely again needs managing. The comparison has been drawn between driving a Porsche at home and a Lada at work, highlighting the tension between the high expectations and experience of some, and the need for universal and affordable access for all. The general base from which we start is low, and thus without revolution and significant investment the result will always trail the expectations and thus be disappointing. We need to find ways to enthuse our audience about the generality of a whole system rather than the extraordinary situation of one 'high-tech', highly specialised application of telemedicine.

Technology of itself should not be a driver of change. Rather, it should enable better outcomes to be achieved from a redesigned process. This involves changes in behaviour, some of which are ingrained within the individual or institutionalised within the organisation. Both need to be challenged, and boundaries between organisations, professions and individuals require renegotiation.

Education and development

General and non-specific telemedicine offers enormous challenges and opportunities. Through education and development we must learn how to identify more accurately where technology can enhance a required outcome in terms of clinical practice and benefit to the patient, and how we can redefine or redesign those processes.

At a more practical and fundamental level we need to improve computer literacy and develop Web-browsing skills. Data capture, storage and retrieval need to be enhanced and distinguished from information analysis. This in turn needs to support clinical governance and critical incident analysis. All of these aims are easy to cite and notoriously difficult to achieve.

Successful implementation

When planning for a successful implementation for a natural community-wide telematics project, consideration should be given to achieving the following:

- shared vision and values
- resolution of outstanding policy issues
- consideration of the whole system
- focused research
- co-ordinated development
- agreed standards
- consistent resource
- a learning style
- supportive education, training and development programmes
- community involvement
- doing what you say you will do, and evaluating afterwards.

An evidence-based and pragmatic start is essential, as the first implementation will have a major influence on future projects. An initial modest success, proven by objective evaluation, will create confidence and a desire to build further. An initial failure (actual or perceived), ongoing tensions or overrunning time and budget will make any further progress significantly more difficult to achieve. Unfortunately, there is little available evidence or experience to help commissioners to move forward successfully.

References

1 Robins S (1999) The pioneer's tale. In: M Rigby (ed) *Realising the Fundamental Role of Information in Health Care Delivery and Management.* The Nuffield Trust, London.
2 NHS Information Management Group (1995) *The Handbook of Information Security.* NHS Information Management Centre, Birmingham.
3 Department of Health (1997) *Report of the Review of Patient-Identifiable Data (Caldicott Report).* Department of Health, London.

The views expressed in this chapter are those of the author, and do not necessarily represent the view of Tees Health Authority.

CHAPTER 14

Telemedicine: what are the benefits for patients and are their interests represented?

Frances Presley

Monitoring changes in the health service is a key function of Community Health Councils (CHCs), and telemedicine is a new and 'radical alteration'.[1] However, despite their statutory right to be consulted, CHCs have had very little involvement in this important debate.

Operating with only a handful of paid staff (two or three full-time equivalents, and often even less in Wales), and with members offering their time for no remuneration, CHCs have provided an extremely low-cost health service 'watchdog' since 1974. However, it has not all been plain sailing, and CHCs are not without their critics, but as the statutory representative of their communities they have proved resilient and are one of the few bodies within the NHS that are still recognisable following the reforms of the NHS during the 1980s and 1990s.[2]

Community Health Councils as the local voice of patients

Founded in 1974 following a series of scandals involving long-stay hospitals, CHCs were viewed as a link between the NHS and the community that they served, separating the management of service provision from the representation of patient and community interests. 'Consumerism' was a rising force, and the need to give voluntary groups a voice was acknowledged by allowing voluntary bodies to elect representatives. At that time it was also considered important for there to be more local authority input into the NHS. Hence local authority representatives also made up the membership of the newly created CHCs.[3]

There are 207 CHCs in England and Wales (16 Health Councils in Scotland and four Health and Social Services Councils in Northern Ireland perform similar functions to CHCs). Each CHC has around 16 to 30 members, half of whom are local authority nominees, a third of whom are elected by the local voluntary sector, and a sixth of whom are appointed by the Secretary of State for Health (or the Secretary of State for Wales for Welsh CHCs). CHCs are funded from a national budget held by the NHS Executive, but they are independent of the NHS management structure, each other and the Association of Community Health Councils for England and Wales (ACHCEW).

Health authorities are required to consult formally with CHCs on substantial variations in service provision, to provide information required by CHCs in carrying out their public duties, and to arrange a meeting between the authority and the CHC members once a year.

The main roles performed by CHCs include the following:

- monitoring local service delivery (CHCs are able to inspect NHS premises, but not the premises of general practitioners, dental surgeries and other non-NHS owned premises unless this has been written into the service provider's contract)
- representing the public and providing their communities' views during consultative exercises
- offering advice and assistance to individuals (this usually includes offering advice and assistance when individuals wish to complain, although this is not an explicit statutory requirement).

ACHCEW: the national forum for CHCs

The Association of Community Health Councils for England and Wales (ACHCEW) was set up in 1977, under provisions of the NHS (Reorganisation) Act 1977, to provide a forum for member CHCs, to provide information and advisory services to CHCs, and to represent the user of health services at a national level. CHCs are not obliged to be members of ACHCEW but the overwhelming majority of them are. CHCs pay an annual subscription based on their own annual budgets and ACHCEW's Annual General Meeting decides national CHC policy.

ACHCEW's statutory duties are as follows:

- to advise CHCs with regard to the performance of their functions
- to assist CHCs in the performance of their functions

- to represent those interests in the health service which CHCs are bound to represent.

Each year ACHCEW publishes 10 newsletters, *Community Health Council News*, and 10 briefing papers, *Health Perspectives*, which examine emerging themes and issues within the health service. Also published are occasional papers, *Health News Briefings*, which are often based on surveys of member CHCs. ACHCEW also hosts seminars and conferences and provides a wide range of training courses for CHC members.

Telemedicine and CHCs

When we received an invitation to speak at the conference on tele-medicine, there was general agreement that no one at ACHCEW had any expertise on this subject, nor had we been approached about it formally previously. A search through CHC reports and literature revealed that there was no evidence of any CHCs addressing this issue.

Lack of consultation

We therefore need to ask what stage the debate has reached in the NHS, and why it has not involved patient representatives, such as CHCs. It is surprising to find, given that telemedicine is a significantly new way of delivering services, with apparent benefits to patients, and that it features strongly in the new information strategy, that there has been no national-level consultation with the Association.

It is clear that substituting a hands-on service with a remote consultation service is a change of service falling within the consultation requirements as described above. Equally, the future loss of a telemedicine service would be the equivalent of a closure, again requiring consultation.

The effect on CHCs' advisory role to patients

Normally we expect patients with a problem or complaint to contact the CHC where they live. That CHC will be familiar with the local health services and will be monitoring them. However, a telemedicine service could pose an additional complication when the service is provided at a distance, with issues of provider responsibility. This can create difficulties when, for example, a patient is referred out of the area to a specialist hospital. In a recent case both the local hospital and the specialist hospital refused to take responsibility for transport arrangements. This makes the work of the local CHC more complicated, as well as requiring clarification of referral responsibilities. This would be particularly important if the link was with another country where there are different medico-legal requirements.

A quick survey of CHCs and telemedicine

As a result of this invitation I read some of the literature on telemedicine and I also conducted a quick nation-wide survey of CHCs. Partly due to time pressures, I sent out a fax to all 207 CHCs in England and Wales asking them to contact me if they had any views on or knowledge of telemedicine.

I gave a brief explanation of what telemedicine is, and emphasised its role within the NHS Information Strategy. This was followed by a series of questions, based on my own preliminary research.

- Who is responsible for the service?
- Will it strengthen or destabilise local services?
- What are the possible dangers of removing the service to a remote site?
- Are there problems of confidentiality and data protection?
- What is the effect on communication between doctor and patient?
- Is it used in rural areas?
- What are the travel benefits?
- Is there access to expert diagnosis?
- Is it finance driven?

I only received 15 replies from CHCs. Four of these had no knowledge or experience, but two were interested in the debate. Four others knew of perhaps one person in the CHC who had some knowledge of or interest in telemedicine, and referred to some of the issues I had raised. Six referred to specific examples of telemedicine in their area. These were as follows:

- Ceredigion CHC – Ceredigion and Mid Wales NHS Trust
- Cornwall CHC – Cornwall Healthcare Trust
- Isle of Wight CHC – St Mary's Hospital (Queen Charlotte link)
- Hammersmith and Fulham CHC – Parsons Green (Belfast link)
- Islington CHC – Whittington Hospital
- Wirral CHC – Arrowe Park Hospital.

Issues for patients and CHCs in telemedicine services

Rapid access to care

From the survey it was clear that telemedicine can provide rapid access to more specialised levels of care which would not otherwise be available, as well as avoiding unnecessary journeys for sick patients.

Dr Monica Williams, Chief Officer for Ceredigion CHC in Aberystwyth, gave a talk on telemedicine at Cardiff in April 1998, and she used the example of cancer services in Wales. She commented favourably on improved access to a range of healthcare services, and how it can avoid the need for patients to make difficult journeys in remote rural areas.[4]

Another response came from a CHC officer who had an interview with the people who are developing telemedicine for British Telecom (BT). BT have been installing pilot projects and they have been very successful. For instance, on the Isle of Wight gynaecologists have used an ultrasound reading link between the hospital and Queen Charlotte hospital in London to scan for abnormalities and provide expert advice.

Kidderminster CHC has provided other examples of services of which they are aware. These include links between CT scanners in district general hospitals and neurological units, links between GPs, hospitals and paramedic ambulance crews and the local cardiac unit for transmitting ECGs, interdepartmental links within a large hospital, and links between the hospital and local GP surgeries. Kidderminster CHC particularly emphasised the benefits of transferring electronic patient records and phosphor-plate X-rays. However, as yet they are not aware of any sites with comprehensive electronic patient records.

Funding

Once BT's pilot on the Isle of Wight had been completed there was no clear continuity of funding from the health service. Similarly, Islington CHC are aware of a dermatology telemedicine project at the Whittington Hospital, although they have no direct experience of it. It has been positively promoted and has a high profile, but there is a negative message about funding. If telemedicine is not being planned and funded properly over a long-term time scale, it is not surprising that CHCs and other patient representatives will remain uninterested and sceptical about it.

Minor injuries units

One successful long-term project, brought to our attention by Cornwall CHC, has been set up at Cornwall Healthcare Trust in a minor injury unit. The community trust buys services from the district general hospital. This means that in a rural location the patients do not have to travel so much, patients can be treated more quickly, and they have the reassurance of being able to see a doctor. The nurses are trained by the Accident and Emergency department. The project has been closely checked and visited, and they have clear referral procedures and protocols. The CHC considers that it is an excellent project and members have seen it in action.

In contrast, in Fulham a minor injuries unit was connected to a Belfast hospital by telemedicine. The local CHC considers this to have been a mistake, given that there were suitable local acute hospitals which could have provided a similar service. It is vital to have clear referral procedures and protocols between the outlying unit and the acute trust in which, if necessary, immediate transfer arrangements can be made for a patient. Fulham CHC believes that in cities it is desirable for the specialist unit to be as close as possible. However, some of the published evidence suggests that the telemedicine link from Belfast to London was successful.[5]

There have also been questions raised as to whether the service in Fulham was appropriate at all for a minor injuries unit, although again this is contradicted by published evidence. Fulham CHC understands that the local Trust is now looking at a telemedicine service for ailments that are easily treatable or which depend on visual diagnosis alone, such as dermatology. However, the experience in Cornwall suggests that it can work for minor injuries, especially in a rural area and when linked to the local main referral centre.

Other issues which appear to be important from the CHC viewpoint are listed below.

Access to patient records

Patients who approach CHCs are often experiencing difficulty in accessing their medical records or are being asked to pay exorbitant charges. This is a particular problem in general practice, and could possibly be exacerbated by telemedicine, if audiovisual records are kept, but are not accessible.[6] The Trust in Cornwall does not record consultations.

Confidentiality: controls and safeguards regarding patient records

ACHCEW is concerned that the recommendations in the Caldicott Report do not go far enough in protecting patient records during the development of new IT systems.[7,8] The Trust in Cornwall found it necessary to reassure patients that their image would only be seen by the doctor concerned and would not be broadcast more widely. Monica Williams, from Ceredigion CHC, also points out that 'without satisfactory arrangements the very advantages of computerisation, relating to ease of access, storage and processing, could severely compromise the confidentiality of patients' records'.[4]

Remote communication

Will telemedicine prove depersonalising and will patients feel inhibited about asking questions? Healthcare professionals will need to be fully trained in communication issues. Some CHCs are concerned that there is no substitute for a hands-on service, and they are worried that the service will become increasingly remote. These issues have also been raised with respect to NHS Direct, the government's nurse-led telephone helpline, although NHS Direct does have clear protocols for calling in expert advice. In the words of Kidderminster CHC, 'It must not be forgotten that even in the next millennium the specialist's history obtained across a desk, and his hands-on examination of the patient, must remain the cornerstone of diagnosis and hence treatment for probably the majority of patients who are referred by their GP for a second opinion.' Monica Williams is more positive about the technology and patients' reaction to it, although she believes that people should still have the choice of more conventional access to treatment if they so wish.

Finance

Is it finance driven in a negative sense? This does seem to be the case from my reading of some of the American literature. For instance, in the Texas Telemedicine Project it is being used as a solution to a problem which should be tackled in terms of the overall funding of the health service.[9] Kidderminster CHC believes that health authorities in financial difficulties are already advancing these untried solutions as a justification for closure or downgrading of small acute general hospitals.

Audits

It is crucial that all developments in this field that directly affect the method of delivery of patient care are subjected to a formal controlled trial. Such trials must be assessed and audited by independent patient and professional groups.

Conclusion

This chapter represents a starting position for the Association of Community Health Councils in addressing the issues of telemedicine and its benefits and disadvantages to patients. On the whole there are clear benefits to be gained from telemedicine, such as rapid access to more specialised care, and the avoidance of unnecessary travel. However, there are possible dangers in terms of provider responsibilities, communication difficulties, referral procedures, issues of confidentiality and access to records. The national statutory body which represents the public interest in the health service should be consulted on both the principles and the specific instances of this major change in the pattern of service delivery, especially where it concerns the clinician–patient relationship. It is difficult to believe that such open debate has not occurred so far, and that so little evidence is yet available with regard to the effects on patients and overall service delivery.

References

1   Wootton R and Darkins A (1997) Telemedicine and the doctor–patient relationship. *J R Coll Physicians London.* **31**: 599.
2   Fereday G (1999) Community health councils: helping patients through the complaints procedure. In: M Rosenthal *et al.* (eds) *Medical Mishaps.* Open University Press, Buckingham.
3   Hogg C (1986) *The Public and the NHS.* ACHCEW, London.
4   Williams M (1998) Paper presented at Telemedicine Conference in Cardiff, April 1998. Available from Ceredigion CHC, Aberystwyth.
5   Darkins A, Dearden CH, Rocke LH *et al.* (1996) An evaluation of telemedical support for a minor treatment centre. *J Telemed Telecare.* **2**: 93–9.
6   Association of Community Health Councils for England and Wales (1998) *Medical Records: Restricted Access, Limited Use.* Health Survey No. 1. ACHCEW, London.
7   Association of Community Health Councils for England and Wales (1998) *Submission on the Caldicott Report.* ACHCEW, London.
8   Association of Community Health Councils for England and Wales (1996) *Health Perspective on Confidentiality in the NHS.* ACHCEW, London.
9   Harris LM (1995) *Health and the New Media.* Erlbaum, Mahwah, NJ.

CHAPTER 15

Identifying the European legal and ethical issues of emergent health telematics

Petra Wilson

Introduction

The title of this chapter potentially promises a wide range of issues, so the first job of its author is to make a loud declaration of *caveat emptor*. Even if the author were capable of outlining every possible legal issue that a European practitioner of telemedicine might encounter in his or her professional life, such an overview would require a whole volume, not merely one chapter. Thus this chapter has two objectives. First, it is intended to provide a road map of major legal issues which pertain to the practice of telemedicine in Europe and the tools with which to obtain further information about them – that is, an outline of the key terms, key legislative instruments and main sources of such instruments. Secondly, it aims to make apparent to the practitioner that the central legal issues in telemedicine are not far removed from the legal and ethical issues which arise in the practice of medicine generally, and thus to reassure him or her that he or she is in fact already familiar with many of the issues.

Since the time of Hippocrates it has been accepted that three ethical principles form the basis of good medical practice, namely respect for *autonomy* of both the patient and the practitioner, respect for *justice* in the allocation of medical resources, and the principle of non-maleficence or *primum non nocere*. I shall therefore revisit each of these well-known ethical principles in turn and ask what new demands the use of telematic tools in the provision of healthcare services may have added, and what the legal response to those demands has been at a European level. It should be noted that, for the sake of brevity, this chapter does not consider the wider ethical issues attendant upon the legal responses outlined.

However, before beginning with the principle of autonomy, a further *caveat* is necessary. Although the subject of this chapter is *European* legal and ethical issues, it is important to note that the regulation of healthcare is generally, according to the principle of subsidiarity, in the remit of individual member states. That is, there is almost no health-care law as such at a European level. The EC Treaty only provides for legislation on public health (article 152). However, many of the significant legal issues in health telematics are not, legally speaking, health matters but rather matters concerning the free movement of goods, services, people or capital and consumer protection, for which matters the EC may enact legislation. Thus this chapter will introduce many European Directives which make few direct references to the provision of healthcare, but with which all of those who practise telemedicine should be acquainted. A number of different types of legislation exist at European level. For the purposes of healthcare telematics, most EC level legislation will come in the form of a *Directive*. As a small point of reference for those readers who are not familiar with European legislation, a Directive is binding on all member states to whom it is addressed. The Directive dictates 'the result to be achieved' but shall leave up to each member state the choice of form and method of implementation (article 189 EC). When discussing a Directive, reference is made to the Article of the Directive as well as to the 'recitals', which is the common name for the paragraphs of the preamble to a Directive. Other forms of EC legislation are the *Regulation*, which is binding in its entreaty on all member states and has direct effect, and *Decisions*, which are binding in their entirety on those to whom they are addressed. The important point to note is that a Directive will not generally provide a citizen with a right to bring an action directly against another natural or legal person. However, the citizen does have a right to bring an action against a member state which has not implemented a Directive.

Respect for autonomy

In medical practice the practitioner has always had to tread a careful line between providing the best medical care for the patient and respecting his or her autonomy – that is, acknowledging the patient's right to hold views, make choices and take actions based on his or her personal beliefs without hindrance. The principle has enjoyed particu-lar respect in western medicine where the influence of Kant has been used to argue that 'to violate a person's autonomy is to treat that person merely as a means, that is, in accordance with others' goals without regard to that person's own goals'.[1] Yet the medical practitioner is the keeper of specialist knowledge upon whose skill the patient has called to help him or her make the right decisions to restore his or her health. As a result of this tension, the principle of respect for autonomy has been translated in medical practice into three broad practical concerns, namely giving sufficient information to allow the patient to make informed decisions, promising complete confidentiality for any information given to the medical practitioner, and giving information without prejudice or external pressure.

The fact that the practitioner is treating the patient at a distance, or making use of some telematic tool, does not in any way alter the principle of autonomy. However, the use of such tools may require new ways of meeting the existing duties.

The first and last principles change very little with the introduction of telematic tools, except that the patient must be fully informed about the use of any telematic tool and its implications for the care he or she is given, and that, as with the use of any device, the practitioner should not use a telematic device except for professionally acceptable reasons. However, the way in which the practitioner honours his or her duty to respect confidentiality will need to change with the use of telematic tools.

Respect for autonomy and confidentiality of patients' data

Confidentiality of patient information, cited in both the Hippocratic oath and the International Code of Medical Ethics,[2] requires that the medical practitioner maintains the secrecy of information entrusted to him or her by the patient. When using telematic tools the medical practitioner will thus have to ensure that the medium he or she uses to store information or transmit it to another treating practitioner is safe from those who might try to intercept it. That means that the computer and telecommunications systems used must be secure, that all who handle such information must have a high duty of confidentiality, and that the patient will have the right to verify the information that is held about him or her regardless of the medium on which it is held.

The principles of confidentiality as outlined above are reflected in great detail in various European legal instruments. Of these, Directive 95/46/EC on the Protection of Individuals with Regard to the Processing of Personal Data and on the Free Movement of Such Data, usually referred to simply as the Data Protection Directive, is perhaps the most relevant and best known. The object of the Directive is to harmonise data-protection legislation in the member states in order to facilitate the free movement of goods, services and people. The aim is to remove any objection from one member state that it cannot interact with another because the other requires too much or too little data protection. The Directive gives basic rights of consent, verification and correction to individuals, and duties of adequate information and secure storage to data processors.

The following points highlight the key rights and duties that the Directive introduces for the healthcare environment.

- Data protection applies to any operation or set of operations on personal data.
- Personal data is any data relating to an identified or identifiable natural person (in some member states it will also apply to data of deceased persons).
- The duties are on the data controller (in medicine this will usually be the doctor, although it could be a health authority or other healthcare provider. The controller must ensure that all those for whom he or she has vicarious liability comply with the duty).
- Data must be processed fairly and only used for the specified purposes stated when collected.
- Data must be relevant, adequate and not excessive.
- The data subject (patient) must give unambiguous consent to collection.
- Note that consent is not strictly necessary when data are collected in the vital interests of the data subject, but ordinary medical data (i.e. non-'vital') requires explicit consent.
- Data subjects must have reasonable access to data and be permitted to have errors corrected.
- Adequate security measures must be taken to ensure the security of stored data (in healthcare more detailed guidance on such measures is given by Council of Europe Recommendation No. R (97)5 on the Protection of Medical Data).

As a result of Directive 95/46/EC, respect for privacy in telemedicine will fall upon any natural or legal person who handles information about a patient. This will usually be a medical practitioner or someone in his or her administrative team. To a certain extent it will also fall upon the telecommunications service provider who gathers information about the medical practitioner and his or her patients in the course of supplying a service. The telecommunications service provider must ensure privacy in telecommunications, which while also governed by the requirements of the framework directive outlined above, is further regulated by Directive 97/66/EC on Concerning the Processing of Personal Data and the Protection of Privacy in the Telecommunications Sector. The Directive extends certain privacy rights to legal as well as natural persons (Directive 95/46/EC applies only to natural persons). The Directive applies to data processed in connection with the provision of telecommunication services in public telecommunications networks, particularly via ISDN.

Respect for autonomy and ownership of data

As noted above, the data controller must ensure that the data he or she is holding are relevant, adequate, not excessive, and used only for the purposes stated at the time of collection. However, in the medical setting this duty is interpreted widely, so that further processing for scientific research purposes may be acceptable even if not originally declared to the data subject as long as appropriate care is taken to ensure confidentiality (this point is made explicit in recital 34 of Directive 95/46/EC).

However, such further processing does raise questions about who owns the medical data in question. In all forms of medical and medically related research of both a commercial and non-commercial nature heavy use is made of databases of patient data. Although such databases may have been suitably stripped of identifying features so that patient confidentiality is no longer a key issue, questions will often arise as to who owns the data in the database. It is not possible to enter into a detailed discussion of European intellectual property law here, but only to note that a recent Directive on the Legal Protection of Databases (96/9/EC) has sought to ameliorate the previously weak protection offered in copyright for databases by creating two types of database and two classes of copyright protection. The stronger type of protection would apply where the creator of the database must be able to show that the database 'by reason of the selection or arrangement of their contents, constitute the author's own intellectual creation' (Article 3(1)). It has been argued that the terms 'arrangement' and 'selection' could refer to either the physical grouping of material or the way in which it is presented to the user.[3]

However, many medical databases are important and useful on the basis that the creator has not made any selection – that is, the database is a comprehensive listing of a given phenomenon. In such cases it will be difficult to satisfy the test described above, but the lesser protection of *sui generis* protection to prevent the extraction and re-use of all or a substantial part of the data presented in the database may apply. This right is provided for in Article 7(1), and will accrue to a maker of a database in cases where 'there has been qualitatively and/or quant-itatively a substantial investment in either the obtaining, verification or presentation of the contents'.

The role of electronic documents and signatures in ensuring respect for autonomy

In order to respect the autonomy of the individual, it is important not only to respect the confidentiality of data stored about a patient, but also the integrity and availability of such data. This means that as well as ensuring that unauthorised individuals do not have access to patient data, the data controller must also ensure that any data he or she sends to or receives from another has 'integrity' (i.e. that that which he or she is receiving is exactly what the sender sent). Similarly, the controller must also be able to be sure that the data has really been sent by the individual shown as the sender. In healthcare it is also important to ensure that data are available when needed, as a database of patient data which is inaccessible for significant periods is worse than useless.

One tool that is important for ensuring confidentiality, integrity and authenticity is the electronic signature, with which an electronic document is signed to show the author and to verify that the contents have not been changed. A wide range of electronic signature tools exist, and they have already been in general use for some time. However, legally they are not yet widely recognised, and many member states are only now introducing legislation to accept an electronic signature as equal to a handwritten signature and amending legislation to allow electronically signed documents to assume full parity with paper documents.

A number of factors will speed up the slow acceptance of these measures. Perhaps the most significant of these is that the field of electronic commerce cannot develop to its full potential until citizens feel confident in the security of their financial interests being trans-mitted via the Internet. Such security measures will soon become the reference standard for the secure transmission of health-related data. A further significant factor in the uptake of legal recognition of digital signatures will be the European Commission's proposal for a directive aimed at establishing a legal framework for electronic signatures (COM (1998) 297 final).

The Commission has argued that in order to make best use of the opportunities offered by electronic commerce, a secure environment with respect to electronic authentication is needed. The Commission explanatory memorandum accompanying its proposal for a Directive notes further that electronic signatures are the ideal tool, as they allow the recipient of electronically transmitted data to verify the origin of the data (authentication) and to check that the data are complete and unchanged and thereby safeguard their integrity. Furthermore, the use of electronic signatures allows the recipient to prove the identity of the signatory by introducing a system of certification in which the certification service provider guarantees the identity of a signatory (COM (1998) 297). The proposal seeks to provide the legal framework which will create a homogenous technology-neutral background for the operation of electronic signatures issued through certification service providers throughout Europe.

Although this Directive, if adopted, could remove a significant barrier to the implementation of European telemedicine, it should be noted that it does not provide a basis for the regulation of electronic documents. Therefore more legislative response is required to bring electronic documents on to an equal footing with paper documents. Much national legislation still requires medical documents such as prescriptions or bills to be provided on specific paper forms, and the legal recognition of an electronic signature is not very helpful if the document in question is required by law to be a paper document. It should also be noted that the proposed Directive does not cover issues related to the conclusion of contracts or other non-contractual acts (Article 1). Thus member states are not required to eliminate legal rules that demand paper forms (moreover, there is generally no EU competency to require changes in such legal formalities).

It should be noted that the legal issues outlined in this section will fall primarily on the medical practitioner (or the healthcare service administration) both for his or her own actions and for the actions of any staff for whom he or she may be vicariously liable. In general duties of confidentiality, integrity and availability will not fall upon the telecommunications service provider. Since it is argued that in general the telecommunications service provider must supply services at reasonable costs, increasing their liability to a broad nature would lead to a raising of costs. It is argued that consumers' interest in low costs takes priority over consumers' rights to compensation.

Respect for justice

In medico-legal terms, ensuring justice in resource allocation is probably the most difficult of the key ethical duties to obey.[4,5] Much lies outside the control of an individual doctor or hospital, depending on national policies for rationing a resource which is always scarce.

Telemedicine is, of course, in itself a means by which some of the injustices of medical resource allocation can be overcome. The patient living in a remote area need no longer be denied access to state-of-the-art medical advice simply because he or she is physically far removed from a hospital which houses the required expertise (in both human and technological terms).

As might be expected, there is relatively little European-level legal activity which concerns the ethical principle of justice in telemedicine, not least because the politics of medical resource rationing are clearly a member states issue. European legislation would only have a role to play if a particular group in society was being unfairly treated on the basis of race, gender, ethnicity or religion or if the way in which a service was being provided was anti-competitive at a European level.

Generally, the respect for justice will be reinforced legally by member state legislation which will ensure that the weaker party (the patient) in any medical contract is duly protected and that the fiduciary relationship of care between the practitioner and patient is legally reinforced so that the patient will be able to gain compensation if he or she has suffered injustice at the hands of the medical profession or medical care system. It is important to note here also that if a more just allocation of healthcare resources can be achieved through the use of telematic devices or telemedicine, a patient might have a cause for complaint if a healthcare provider was failing to use such a system.

Justice and the distant consumer

In a traditional telemedicine situation, the patient or citizen and the healthcare practitioner will never meet face to face. Sometimes this relationship will be mediated by a practitioner who is meeting physically with the patient, but at other times the link will be direct. In such a situation the citizen will be contracting at a distance with the healthcare provider, who will in turn be required to comply with his or her duties under Directive 97/7/EC on the Protection of Consumers in Respect of Distance Contracts.

In Directive 97/7/EC, the European Parliament and Council have sought to provide special protection for the consumer from the special risks which may arise when the consumer is unable, because of distance, to examine the goods before purchase or to see the service supplier's premises. The Directive will apply if the consumer is a natural person acting outside his or her professional capacity (i.e. as a citizen) and if the supplier is acting in a commercial or professional capacity. It will apply to any situation in which the supplier makes use of one or more means of distance communication in order to conclude the contract.

Thus, if a citizen contracts with a telemedicine practitioner via the Internet, for example, the telemedicine practitioner will have to ensure that he or she has fulfilled his or her duties under the Directive. On the whole these are not very onerous duties, and they include the consumer's right to written information about the supplier and the goods or service and the right to withdraw from the contract within 7 working days after the time at which the written information is supplied. Although these requirements will generally apply to the supply of medical services, it should be noted that emergency situations may arise in which the consumer is incapable of receiving such information or exercising his or her right to withdraw. In such cases the usual legal standards of tortious liability would be applied to assist the consumer, should some harm have occurred.

Justice and Universal Service Obligation

The political and legal concept of Universal Service Obligation (USO), by which governments seek to ensure that equal and fair access to essential services is available to the whole of society, regardless of geographical location, is also tangentially relevant to the supply of telemedical services. A significant action in this issue at EU level was the European Parliament and the Council Directive 98/10/EC on the application of open network provision (ONP) to voice telephony and on universal service for telecommunications in a competitive environment (replacing European Parliament and Council Directive 95/62/ European Community).

The objective of the ONP directive was to take account of the liberalisation of the telecommunications market by 1 January 1998 and to guarantee provision of a defined universal service for telecommunications in the European Union. Among many demands, the Directive makes the following key demands of each member state.

- Member states shall ensure that the telecommunications organisations provide a fixed public telephone network and a voice telephony service. This will include public pay-phones and telephone terminals suitably adapted for the elderly, disabled and those with special social needs.
- They shall ensure that users can, on request, obtain a connection to

the fixed public telephone network and connection and use of approved terminal equipment situated on the user's premises, in accordance with national and Community law. Such terminals will support international calls, fax and/or data communication.

Primum non nocere – protecting the patient from harm

The third grand principle of medical ethics is that the healthcare provider shall attempt to benefit the patient or at least shall not harm him or her. The role of law in this duty has principally been to assist the citizen who has suffered harm, thereby also acting as a warning to the careless practitioner of the possible consequences of his or her actions. Thus key questions for the provider of health telematic services are always 'What am I liable for?', 'What must I ensure?' and 'What are my minimum duties?'

There is no European-level legislation which directly states the liability of the telemedicine practitioner. At member state level specific legislation, as well as the law of contract and the law of negligence (tort/ delict), will govern the relationship between practitioner and patient so that the patient can claim compensation when injured as a result of a telemedical treatment. However, two directives in the area of product liability may also be important.

Causing harm through defective products

The EU Product Liability Directive (Council Directive 85/374/EEC) imposes a duty of strict liability on producers. In order to establish liability, two elements must be present, namely a defect in the product and harm to the consumer. Whether the defect is the result of negligence on the part of the producer is immaterial.

The scope of the Directive is very wide, covering all constituent parts of movable products, including electricity and intellectual property such as software when it forms an integral part of the operational product.[6] In seeking to provide good protection for all citizens, the Directive further stipulates that when the producer cannot be identified, liability will lie with the supplier of products. Thus a patient who is injured by a defective telematic product would have cause for action against the producer (if identifiable) or the medical practitioner. The medical practitioner, being also a consumer, could also have an action against the supplier or producer from whom he or she acquired the product if an injury arose through the use of a defective product used in a reasonable and responsible way. As a result of the duties imposed on producers by the Directive, the need for product testing, quality control and risk monitoring cannot be over-emphasised.

Causing harm through medical devices

A supplement to the general product liability legislation is found in Council Directive 93/42/EEC on Medical Devices, according to which a producer of a medical device must comply with the standards set by the European Committee on Standardisation (CEN) and the European Committee for Electrotechnical Standardisation (CENELEC). According to the Directive, a wide range of medical products will be covered, including :

'any instrument, apparatus, appliance, material or other article, whether used alone or in combination, including software necessary for its proper application, intended by the manufacturer to be used for human beings for the purposes of diagnosis, prevention, monitoring, treatment or the alleviation of disease, diagnosis, monitoring, treatment or the alleviation of or compensation for injury or handicap, investigation, replacement or modification of the anatomy or of a physiological process, or control of conception.'

Although the Directive will not of course cover all telematic devices, as it is restricted to those devices which are in immediate contact with the human body, a number of products – such as medical imaging devices – will come under its scope. Accordingly, any producer who does not meet the standards required by CEN or CENELEC will incur liability.

Harm arising from non-use of telemedicine

Until now we have stressed the possible legal consequences arising from the use of telemedicine. However, it should be noted that as telemedicine becomes more common, and indeed the norm in certain areas of medical practice, those practitioners who fail to use telemedicine might be regarded as negligent should the patient be able to show that harm or injury had befallen them which would probably not have occurred had telemedicine been used.

Conclusion

It may be argued, on the basis of the preceding pages, that the growth of telemedicine should not bring with it a medical litigation explosion. On the whole, the legal ramifications of the interaction between the healthcare provider (whether as healthcare authority or as practitioner) and the patient will change very little. Although new players will enter the medico-legal arena, notably the telecommunications provider and, in time, the added value telecommunications service provider (such as the Trusted Third Party), they will only occasionally enter the court-room.

However, the key duties of respect of confidentiality, fair allocation of resources and avoiding harm will continue to dominate the interactions between medicine and patient in telemedicine as in face-to-face medicine. The practitioner of telemedicine must therefore remain aware of his or her duties and must examine each new procedure to ensure that he or she is able to maintain the same high standard of care in telemedicine as in any other specialism.

References

1   Beauchamp T and Childress J (1994) *Principles of Biomedical Ethics.* Oxford University Press, Oxford.
2   Mason JK and McCall Smith A (1991) *Law and Medical Ethics.* Butterworths, London.
3   Cerina P (1993) The originality requirement in the protection of databases in Europe and the US. *Int Law Rev Industrial Property Copyright Law.* **591**.
4   Sharpe E (1981) *Justice and Health Care.* D. Reidel, Boston, MA.
5   Newdick C (1995) *Who Should We Treat?* Oxford University Press, Oxford.
6   Howells G and Weatherill S (1995) *Consumer Protection Law.* Dartmouth Publishing, Aldershot.

Further reading

European and international data protection instruments

A great deal of literature exists on national, European and international data protection instruments. For detailed coverage with frequent updates see:

Charlton S *et al.* (1996) *Encyclopaedia of Data Protection.* Sweet and Maxwell, London (regular updates as necessary).

Simitis S *et al.* (1996) *Data Protection in the European Community.* Nomos, Baden-Baden (twice-yearly updates).

Security in healthcare computing

For an introductory overview of these simple security issues see:

Barber B, Treacher A and Louwerse K (1996) *Towards Security in Medical Telematics.* IOS Press, Oxford.
Benson T and Neame R (1994) *Healthcare Computing.* Longman Group, London.
van Bemmel JH and Musen MA (1997) *Medical Informatics.* Springer Verlag, Heidelberg (Chapter 33 on security in medical information systems).

Legislation

Directive 98/10/EC on the application of open network provision (ONP).
COM (1998) 297 – Commission's proposal for a directive aimed at establishing a legal framework for electronic signatures.
Directive 97/7/EC on the protection of consumers in respect of distance contracts.
Directive 97/66/EC concerning the processing of personal data and the protection of privacy in the telecommunications sector.
Council of Europe Recommendation No. R (97)5 on the protection of medical data.
Directive 96/9/EC on the legal protection of databases
Directive 95/46/EC on the protection of individuals with regard to the processing of personal data and on the free movement of such data.
Directive 93/42/EEC on medical devices.
Directive 85/374/EEC on product liability.

CHAPTER 16

Key issues in taking health telematics safely and ethically into the 21st century

Michael Rigby, Katherine Birch and Ruth Roberts

The preceding chapters of this book have illustrated what telematics in health can achieve, as well as highlighting a range of issues that need to be considered when healthcare harnesses telecommunications. However, telematics in health is not merely about automating previous processes, although it is important for healthcare's core values to be maintained. Moreover, it would be seriously misguided to assume that the public, professionals or organisations will continue past patterns of behaviour when working in new ways in a communications-based culture.

It is safe to assume that telematics and telemedicine will grow rapidly to become major contributors to healthcare delivery. It is therefore important to understand and address the new issues that they will bring not only at a local level but also globally, as telecommunications are rapidly breaking down traditional boundaries and constraints. This chapter seeks to identify these issues.

Effects on clinical practice

The first important area to consider concerns the effects on clinical practice. Healthcare structures, from principles of practice to training structures, have evolved over time to serve a purpose within a specific healthcare setting. Telematic and telemedicine applications are intended to overcome specific constraints and limitations in that environment, but in changing the old order it is important to consider whether destabilisation or other adverse effects may result from disturbing the previous processes and organisation. Some of the key issues are raised here.

Overloading of experts

In a traditional setting, patterns of referral up the hierarchy of expertise are clear, if somewhat laborious. The time and cost effects of distance provide an effective, if imperfect, means of ensuring that referrals to centres of excellence or individual experts beyond the local primary–secondary care hierarchy are restricted to those with exceptional clinical needs and to the exceptionally determined. Telemedicine demolishes this control mechanism at a stroke. As telemedicine systems become more widespread, national or international centres of excellence will rapidly come under pressure to take more and more referrals. If they allow themselves to be swamped, they will jeopardise their quality of service, and also individual clinical experts will be more likely to 'burn out' at an early stage.

Destabilising clinical skills development

If telemedicine undermines the primary–secondary–tertiary referral hierarchy, it will also have an adverse effect on clinical skills development, particularly in medical specialities. The present arrangements provide broad strata, with different types of case and complexity being handled at the different levels up a progressive hierarchy, and therefore clinicians seeking to develop their skills and experience can progress through the same hierarchy. However, if the effects of telemedicine are to move a significant amount of intermediate and advanced work to more remote sites, this informal structure of learning opportunities is likely to be destabilised as more local providers reduce their range of activity. The dominant role of more remote centres of excellence would be reinforced, albeit unintentionally, far beyond that intended to produce planned regional expert centres.

Clinical culture development

There is also a need for a change in clinical culture from one of independence to one of interdependence, based partly on the move towards a holistic care approach, but in this context recognising the new interdependent pattern of working which telematically based healthcare necessitates.[1] This new approach needs to be risk aware, not risk averse, and at the same time it needs to break down traditional barriers within and between professions. As the contributors to care, and the evidence on which they draw, become increasingly dispersed, so there is a greater need to ensure a holistic picture of the patient and vision of how their health support should progress.[2]

Clinical and cultural relevance

Care always needs to be appropriate to the patient's setting and circumstances. Even with traditional delivery mechanisms, it is often necessary to adjust initially recommended patterns of treatment or care to allow for the patient's social and other circumstances. Teleconsultation brings specialist expertise to the locality, but the remote expert may be less aware of or sensitive to local practical and cultural issues and to the local care support infrastructure, particularly if the link is from another region or country. Other factors may also come into play. For instance, a remote reference database of specialist pathology values may be based upon a different ethnic and physiological mix of population. Mutual briefing and understanding between the referring and expert clinicians are important to ensure local relevance, feasibility and acceptability of the treatment offered.

Validating remote learning

Telematic facilities provide important opportunities for education and learning (both informal and formal). However, educational providers and professional bodies have invested much time and effort over past decades in devising structured means of ensuring that education and training processes are carefully planned, and that learning has been appropriately acquired and assimilated. This has been done not out of self-interest, but to ensure that clinicians practise effectively with sound use of knowledge and appropriate application of skills. It is therefore important to ensure that remote learning does not circumvent these quality measures which are intended to protect the citizen and ensure good care.

Informal reskilling

A perceived added benefit of some forms of telemedicine, particularly teleconsultation, is that they enable the clinician to update and maintain his or her skills through being an observer and participating assistant in the consultation between the patient and the remote expert. Thus, after referring a number of similar cases, clinicians may feel that they have the competence and skills to undertake the treatment themselves. There is informal evidence, as indicated in earlier chapters of this book, that this is indeed the case for doctors in areas such as teledermatology, and for nurses in minor injuries units connected by a telemedicine link to a major trauma centre, where telemedicine referrals have tailed off as clinical expertise has grown. In this respect, telemedicine is yielding important benefits in improving health service skills and structure, albeit with the result that fixed telemedicine links may be used less frequently than when they were first installed. However, healthcare organisations and the professions have not yet addressed the issue of peer review and clinical audit for these remote services, and there is the risk of skills drift among isolated clinicians who have upgraded their competencies informally in this way.

Validating distant learning

Although improving clinical skills remotely can be perceived as a benefit, it may be difficult to validate the acquired skill independently. Informal side-by-side transfer of clinical skills, sometimes colloquially referred to as 'sitting next to Nellie', is also largely opportunistic but does have the significant advantage of a two-way dialogue. Remote learning (whether informal or formal) needs to be properly structured and evaluated if 'learning from the "Tele"' is to be as sound as 'learning from Nellie', or it needs to have an identified degree of validation as in the case of formal course learning, if inappropriate skills drift is to be avoided.

Record-keeping

The issue of record-keeping can also be adjudicated according to the issues of clinical responsibility and liability. If the local clinician takes external advice and then proceeds with an informed pattern of treatment, the important element is a note in the records of the external advice received. If the remote expert undertakes the treatment, he or she will need to create a record of the teleconsultation or other form of remote history-taking, and retain that record as his or her clinical evidence.

Concepts of clinical records

As clinicians practise remotely, and as scans, video clips and other non-text clinical messages are transmitted electronically, so it will become necessary for these to be recognised as part of the patient's record. New understandings, structures and cataloguing, subject access and controls for the new paradigm electronic patient record need to be devised, whether the record is located with one health provider or distributed in virtual form among many clinicians, possibly in several countries.

Credibility of clinical records

In the manuscript era, with its related localisation of the clinical team, the clinical record could generally be trusted within its use environment because it was built up by members of a distinct clinical community, with the authorship of each entry being identified. By contrast, the electronic record will be built up by a virtual community of clinicians who may be spread over a much wider area, and who may not necessarily be known to one another. Electronic techniques such as automatic identification of authorship and electronic signatures, when adequately employed, can give attribution for purposes of audit and accountability, but this will provide little guidance on the clinical decision-making processes and criteria of a remote clinical colleague.

Perceptions of privacy

The confidentiality of clinical records is well understood and respected in a paper record environment, even if the latter is prone to system failures. The early stages of the systematic introduction of electronic-record networking led to major professional concerns,[3] but full electronic records have the potential for a new generation of issues. For instance, named third parties (whether they be health professionals or family carers) become data subjects in their own right, linking of family members can yield clinical rewards, but with an inadequately debated challenge to confidentiality, and traditional data protection concepts of agreement and consent have not generally been adequately thought through with regard to the situation of health service clients whose rational decision making is temporarily impaired through anxiety or illness.[4,5] There may be direct conflicts of interest which raise new ethical dilemmas – for instance, the benefit to an individual of new tools to calculate hereditary genetic risks, balanced against the challenge to privacy of related individuals (not least through posing questions over paternity).

Potential false goals

In the right settings and applications, bringing the 'tele' into healthcare clearly has advantages, and the examples and analyses cited in this volume are intended to aid decision-makers in reaching the appropriate decisions. However, there are also false assumptions and other pitfalls that need to be guarded against. Some of the more significant ones are introduced here.

Better use of resources

One frequent motivation for applying telematics is that it is assumed to improve resource utilisation. Although that may be the case, allowance must be made for the additional technical resources needed. With regard to health professional time, whilst telemedicine support may enable an appropriate nurse or primary care physician to undertake work which would otherwise be referred to a doctor or consultant, respectively, when teleconsultations are undertaken the fact that there are usually clinicians involved at both ends of the link is often over-looked, thus doubling the clinician involvement compared to a face-to-face patient consultation.

Faster treatment

One argument in favour of telemedicine referrals is that they are a faster route to obtaining an expert opinion. However, the remote consultant does not have any extra hours created in his or her day by the installation of the telematic link – indeed, a telemedicine consultation may take longer than a face-to-face consultation. The effect is simply that use of technology tends to increase urgency independent of content (i.e. the same effect as is created by e-mail and mobile telephones). Thus earlier consultation through telemedicine rather than via a face-to-face appointment may be merely an insidious form of technology-induced queue-jumping.

Local strengthening or weakening?

A sound objective for telemedicine investment, at least in the first instance, may be the strengthening of the network of services available in a locality, particularly when a hub-and-spoke pattern is planned to bring remote areas of a catchment area closer to their secondary care centre. However, with modern telecommunications there is no reason why the primary care sites have to restrict themselves to their local secondary care service, other than because of ingrained tradition or restrictive regulations. This may not be a bad thing in the interests of providing optimum integrated care, but there is an inherent incompatibility between the objective of strengthening the local referral structure and related interprofessional understanding, and that of creating a wider spread of expert referral centres. The tools intended to strengthen local infrastructures may indeed weaken them, and sound development of principles and protocols is needed in order to obtain the best of both aspects and the worst of neither.

Organisational management

Telemedicine applications present a new challenge to organisational management.[6] Because telecommunications flow invisibly into and out of the organisation, it is much more difficult to identify patterns of clinical working compared to, say, monitoring of traditional referrals.

Booking and billing mechanisms

Teleconsultations and other telematic activities consume expensive clinical time as much as any other consultation, but contractual controls and billing techniques are much less developed. New booking systems need to be devised both for teleconsultation suites themselves, and to ensure that clinician time is used effectively, these booking requirements being exacerbated by the fact that two independent healthcare organisations and practitioners have to be co-ordinated in order to synchronise their activities, as identified in the Swedish study.

Needs-based referrals

Even more challenging is the desirability of co-ordinating referrals, to ensure equity between remote referrers and between electronic, telemedicine and paper-based referrals. If telematics and telemedicine are to be used to eliminate the challenges of distance or of local skills deficits in order to achieve equity in healthcare – a long-term European goal[7] – it is important that the new technology should not itself demolish existing equity control mechanisms. Therefore there is a need for protocols to handle referrals systematically on the basis of need, not on the basis of communication method.

Quality assurance in health telematics

In any healthcare delivery situation, consideration of quality should play an important part, and quality assurance must be built in from the outset. Telematics and telemedicine can have many new dimensions with regard to quality, and the principal ones are indicated below.

Quality of the technology

This must be appropriate for the purpose, but there is a danger of adopting quality standards from more basic applications and using them inappropriately for more advanced projects. The quality of basic telecommunications has improved radically, so that for voice communication poor telephone quality is now unusual, and quality of fax transmission of text is also satisfactory. However, fax provides a good example of quality limitations in that it is still not an appropriate medium for sending detailed diagrams or small-print text.

Television and video transmission provide a good example of ensuring appropriate quality for purpose – a low-specification link that produces jerky images may be satisfactory for low-cost video-conferencing between professionals who use the link regularly, but would be unsatisfactory for remote patient consultations. In turn, systems that are satisfactory for conversational linkage may not have the resolution and the colour accuracy necessary for teledermatology. It is in the technical areas of clinical transmission that it is essential to ensure that the overall telemedicine equipment and linkage are of the appropriate quality for the purpose, otherwise the clinicians at either end will be unable to assess the quality of the images reaching their opposite number.

In Chapter 3 a good practical example (from Sweden) was cited of the importance of considering these issues from user perspectives and with a view to longer-term requirements.

At a more prosaic but practical level, adequate reliable and effective technology is essential to support any form of telematic-based practice.

Inadequate access to evidence bases will result in their not being used. Electronic patient record (EPR) and other paperless systems will fail if there is excessive down-time, a slow system response or queuing for available terminals.

Quality of the information

The telematics era has resulted in much wider access to information through electronic libraries, reference sets and Internet sites. However, Impicciatore and his colleagues have highlighted the contradictions and errors of Internet-site advice available to the public for specific conditions.[8]

Instead of often being short of up-to-date information that is readily available, the clinician (and the average member of the public) will be swamped by information, not all of proven source or quality. It will therefore become increasingly important to be able to appraise the quality and utility of that information, using critical appraisal skills.[9]

Quality of the process

The clinical process

However good the equipment may be, inappropriate use of the process will compromise the overall quality of a telemedicine activity. For a consultation, the expert needs to have an adequate view of the patient, but they also need to be aware of any other factors or constraints, such as a third party who may be in the room with the patient but out of view of the video camera. The active presence of a local clinician may be important for transmitting other information, such as clinical pictures, but must not inhibit the patient's communication with the remote expert. The preparation of the patient for the consultation, and the preparatory briefing of the remote expert, are also important factors contributing to quality.

Software and data-handling process

Much faith has necessarily been placed in the accuracy of the software process, both in handling data and in calculation. Furthermore, checking of software, or even validation of the design and underlying clinical logic, are extremely difficult. However, unqualified dependence on systems which are untested and often unproven in large-scale use can be dangerous. For instance, examples do exist in the health press of incorrect autocalculation of blood serum screening results,[10] and of failed scheduling through population-recall systems.[11] Indeed, the European Commission has now commissioned a pan-European project entitled TEAC-Health (Towards European Accreditation and Certification in Health) to consider the whole issue of validation of health telematics and to identify possible solutions.[12]

Quality of outcome

Unfortunately, but understandably in view of the greater difficulties of measurement, there is less literature on the methods or results of measuring the quality of outcome of telematics-assisted processes. Intrinsically, quality of outcome would consider the view of the patient as to whether the consultation or other process was successful in their own opinion, and whether their treatment had achieved the right balance between external high-quality evidence and personal and localised customisation. For telemedicine, quality measures might include whether they felt that they were able to convey the same relevant information, feelings, and anxieties as they would have done with a conventional clinical activity, and would be related also to the quality effects of avoiding long journeys or being treated far away from home and from family support. The wider extrinsic view of the quality of the outcome would be measured by the overall quality of the treatment received and the net health gain, compared to other approaches.

The need for objective planning

It is important to return to the decision-making processes. First, when decisions are made to use telematics, are these for the right reasons? It is important that investment is only for appropriate, evidence-based reasons. Telematics should not be implemented simply as a copycat response to similar applications being introduced elsewhere, and conversely it should not be opposed for inappropriate reasons in situations where it could improve quality of care or cost-effectiveness. An Economic and Social Research Council study due to commence in Scotland in 1999 should be particularly valuable in this respect.[13] This study will investigate the attitudes and knowledge of decision-makers with regard to telemedicine, in order to determine the factors which have enhanced or inhibited uptake.

Secondly, it is the attitudes, skills and beliefs of clinicians that will determine the degree and manner of the uptake of telemedicine systems once they have been implemented, including referral patterns. For instance, one study from the USA indicates that 50% of all medical information systems failed to be used properly because of staff resistance,[14] and another study showed that only 25% of system abilities or functionality of acute hospital systems was utilised.[15] The Scottish study referred to above will also be important for establishing the attitudes and practice of clinicians for whom telemedicine applications are available.

Management of expectations

As has already been shown, there are many expectations of telemedicine – at patient, clinical, organisational and political levels. Not all of these expectations are realistic or evidence based; some are phenomena of the innovative and piloting stage of telemedicine development, and cannot be sustained within any generalised roll-out. Some of the positive features – such as quick access and on-time appointments – would be unlikely to be sustainable if telemedicine became endemic, while its current success in some settings is in effect being achieved at the expense of patients who go through the traditional referral routes.

Similarly, the Internet, and other telematic services, are potentially very promising, but this can lead to unrealistic expectations. For instance, external database search potential is often restricted by pressures of time, limited availability of facilities, and variability of quality which can be difficult to assess.[12] Electronic record systems and also telecommunications support to remote staff can be hampered by shortage of investment. For example, queuing for terminal access or use of outdated software waste clinical time and do not enable potential benefits to be achieved.

Thus it is important that all parties are realistic in their expectations, and that those who are responsible for developing strategies or implementing applications are realistic when they explain the achievable benefits and any likely limitations or drawbacks. Failure to be realistic at this stage will only result in subsequent dissatisfaction which is likely to damage future credibility and co-operation.

Planning for implementation

History has shown that a totally uncontrolled incremental approach to adoption of new clinical technology is inappropriate without a deeper understanding and response. For example, minimally invasive surgical techniques were shown to have advantages in development sites, but the clinical skills of three-dimensional fine manipulation through remote on-screen observation are significantly different to those needed for direct and tactile open-site surgery. Their uncontrolled dissemination and adaptation was not only economically inefficient, but without structured training it incurred a risk of avoidable iatrogenic adverse outcomes.[16] Similarly, but from an organisational dimension, widespread uptake of day-surgery techniques where appropriate has provided major benefits through reduced costs and reduced hospitalisation of patients, but was only fully effective when the required organisational changes, including staffing, buildings and systems, were recognised and implemented.[17]

Thus health telematics, and particularly telemedicine, need a careful yet positive approach if they are to be applied without adverse effects. Although they usher in important new opportunities, there are also new dimensions to traditional issues of accountability, quality assurance and consumer orientation to be addressed, which are a direct result of the prime characteristic of enabling care delivery across all organisational and national boundaries.

References

1   Purves I (1999) The doctor's tale. In: M Rigby (ed) *Realising the Fundamental Role of Information in Health Care Delivery and Management: reducing the zone of confusion.* The Nuffield Trust, London.

2   Rigby M (1999) *Realising the Fundamental Role of Information in Health Care Delivery and Management: reducing the zone of confusion.* The Nuffield Trust, London.

3   Anderson R (ed) (1997) *Personal medical information – security, engineering and ethics.* Proceedings of the Personal Information Workshop, Cambridge, UK, 21–22 June 1996. Springer-Verlag, Berlin.

4   Rigby M, Hamilton R and Draper R (1998) *Towards an ethical protocol in mental health informatics.* In: B Cesnik, AT McCray and J-R Scherrer (eds) Proceedings of Medinfo 98 Ninth World Congress in Medical Informatics. IOS Press, Amsterdam, 1223–7.

5   Rigby M, Draper R and Hamilton I (1998) The electronic patient record – confidentiality and protection of interests for vulnerable patients. In: PW Moorman, J van der Lei and MA Musen (eds) *Proceedings of IMIA Working Group 17,* Rotterdam, 8–10 October 1998 (*EPRiMP: The International Working Conference on Electronic Patients Records in Medical Practice,* Department of Medical Records, Erasmus University, Rotterdam, 248–52).

6   Rigby M (1999) The management and policy challenges of the globalisation effect of informatics and telemedicine. *Health Policy.* 97–103.

7   Roger-France F, Noothoven van Goor J and Staehr-Johansen K (1994) *Case-Based Telematic Systems Towards Equity in Health Care.* Studies in Health Technology and Informatics Vol. 14. IOS Press, Amsterdam.

8   Impicciatore P, Pandolfini C, Casella N, Bonati M (1997) Reliability of health information for the public on the World Wide Web: systematic survey of advice on managing fever in children at home. *BMJ.* **314**: 1875–9.

9   Roberts R (1999) *Information for Evidence-based Care.* Harnessing Health Information Series. Radcliffe Medical Press, Oxford.

10 Cavalli P (1996) False-negative results in Down's syndrome screening. *Lancet.* **347**: 965–6.

11 Computer error leads to smear recalls failure. *Health Serv J.* **108**: 5603–6.

12 www.multimedica.com/TEAC

13 Ibbotson T, Reid M and Grant A (1998) The diffusion of telemedicine: theory in practice. *J Telemed Telecare.* **4**: 1–3.

14 Anderson JG, Jay SJ, Perry J and Anderson MM (1994) Modifying physician use of a hospital information system – a quasi-experimental study. In: JG Anderson, CE Aydin and SJ Jay (eds) *Evaluating Health Care Information Systems: methods and applications.* Sage, Thousand Oaks, CA, 276–87.

15 Anderson JG and Aydin CE (1994) Overview – theoretical perspectives and methodologies for the evaluation of health care information systems. In: JG Anderson, CE Aydin and SJ Jay (eds) *Evaluating Health Care Information Systems: methods and applications.* Sage, Thousand Oaks, CA, 5–29.

16 Banta HD (1993) *Minimally Invasive Therapy in Five European Countries: diffusion, effectiveness and cost-effectiveness.* Elsevier, Amsterdam.

17 Audit Commission (1990) *A Short Cut to Better Services: day surgery in England and Wales.* HMSO, London.

CHAPTER 17

And into the 21st century: telecommunications and the global clinic

Michael Rigby

This book commenced with an acknowledgement of the virtue in virtuality, and the major potential which modern telecommunications offers to global society, including the health sector. Telematic activity, including teleconsultation, is no longer unusual even though it may not yet be commonplace. However, the potential significance of its effects is not yet fully appreciated.

What was locality?

As an example, local and national healthcare frameworks will seem increasingly outmoded. For instance, reference databases are increasingly strengthened by being compiled internationally (as emphasised in Chapter 4 by Jari Forsström), several European health telematics research projects under the Fourth Framework Telematics Application Programme have explicitly sought to overcome the restrictions of national boundaries, and the example of remote community telemedicine empowerment described from Sweden in Chapter 3 to link to local referral centres could as easily link to an expert location in another country.

The global clinic . . .

Thus partly by accident (such as Internet availability), partly through the impending general accessibility of leading-edge telecommunications technology (such as domestic video-phones), and partly through traditional healthcare pressures for specialisation and rationalisation, the potential has arrived for the public and the practitioner to seek health advice and treatment from anywhere in the world. The global clinic is upon us, and can be entered from a domestic living-room.

. . . and the unknown effects

Earlier chapters of this book, written by experts from across Europe, have shown the considerable benevolent potential of telematics applications in health. However, the young and largely 'one-off' nature of the applications described, and the unresolved issues raised by every contributor, should also trigger caution in the face of widespread populist pressures for rapid expansion of telemedicine and other health telematics systems ahead of further objective study.

Almost all of the contributions highlight three worrying problems. The first is a general underlying assumption that telematics and telemedicine merely involve the modernising through teletechnology of existing clinical techniques and paper-based processes – an assumption that is severely challenged by the analyses. The second is the serious lack of research evidence, particularly that related to behavioural changes and organisational effects. The third problem is the need for a new order of validation and management measures to address the globalisation effect of modern telecommunications on healthcare delivery.

Thus two purportedly citizen-focused and consumer-friendly dynamics appear to be in direct conflict. The general pressure to harness the latest technology to achieve ostensibly better and more convenient services is in principle laudable, and is very much in tune with the latest consumerist vogue. However, in contrast, awareness is also growing of the need to protect the citizen against concealed hazards, and consequently to ensure that healthcare is evidence based. These core principles cannot be assured in telematics without further study of the medium, and objective evaluation methods for the service delivery modes.

There is therefore an urgent need for the healthcare sector and its leading bodies to take action if health telematics is to be taken safely into the next century. The challenge is to reconcile citizens' rapidly changing expectations in the light of the new order in telecommunications, with the essential special characteristics of health and healthcare identified in the opening chapter.

Above all, the effect on healthcare culture and organisations has not yet been considered, yet it will be profound. An early but minor sign of this is the questioning of treatment plans by the citizen who has undertaken hours of Internet searching which the clinician cannot match in an essentially brief consultation.

The arrival of the tele-first century

In all walks of life the public is rapidly adopting and expecting a telecommunications-based society. Television has become a dominant social force, conveying not only constant entertainment and education, but providing a prime source of subjects of general conversation while at the same time subliminally having a major influence on attitudes and expectations, including attitudes to healthcare. Television is likely to become even more powerful in the near future as interactive facilities become available.

The telephone and the virtual organisation are at the root of modern commerce, not least of retail service organisations. The public now expects to be able to undertake major transactions, including private banking, mail-order purchases and booking of hotels and holidays, by telephone. The theoretical inherent risks which would be identified by a rational analysis (ranging from impersonation through to misuse of credit card numbers) are accepted with little anxiety in return for the major benefits of convenience, choice and cost-effectiveness.

Furthermore, the virtual organisation is taking on a deeper meaning as individual organisations aggregate together, based very much on telematics, in order to provide a functioning virtual economic sector. Thus, for example, in air transport competing airlines, individual airports and national air traffic control centres all work in collaboration to provide a seamless service that is only possible through telematics. Moreover, within this the consumer public can, for instance, obtain a real-time update on the progress of an individual flight either by ringing an automated telephone system at the airport, by accessing the airline's Internet site, by checking on broadcast television channels' text-based information systems or by reading the monitors at the airport. Yet this demonstration of operational seamlessness is paradoxically further highlighted by a hidden complexity, in that many of these organisations themselves are largely virtual, consisting of many sub-contractors working together as one entity – indeed often under a single identity.

It's the message which matters

Whereas to McLuhan the medium was the message,[1] now the message is the product. People's overriding demand is to know, and to know

now, and they expect the new medium to deliver, whilst at the same time being oblivious to the technical and organisational issues, and dismissive of attempts to explain the problems.

Today society expects telematics to be applied comprehensively. Whenever a problem arises with a transaction, the immediate consumer response nowadays is 'pick up the telephone and find out'. Within a decade, direct sales telephone-based insurance has moved from being an unlikely innovation to being a standard expectation. New applications, such as Internet shopping, show evidence of being rapidly adopted. Bills are paid and problems dealt with through teleservice centres that operate well beyond the traditional working day, and from geographical locations which are neither known by nor relevant to the consumer. Thus, as the public rapidly adopts and enjoys television culture and telematics-based services, and will no longer accept time or distance as excuses given the universality of telecommunications, we could truly be said to be entering the tele-first century.

Lessons from the past – the rise of the railway age

However, any assumption that this newly empowered and tele-educated population will merely want speeded up old-style services, or that the application of telecommunications will simply modernise paper-based procedures within existing organisations, would be seriously flawed. It would also fail to identify the potential problems – ranging from essential reskilling to new safety hazards – which will inevitably arise. It is therefore important to anticipate and understand the significant changes which are an inevitable consequence of the telecommunications revolution.

Study of the effects of the advent of the railways some 150 years ago provides a useful indication of the organisational and social revolution which may be in store. Although the technology is very different, the fundamental objectives and effects are very similar – the railways took the time out of distance through reducing to a few hours journeys which had hitherto taken days, just as telematics now takes the distance out of transactions, potentially including those in healthcare. Yet the profound and extensive effects of the railways were never anticipated by those first engineers who mounted a steam engine on a moving wagon in order to haul the hitherto horsedrawn carriage.

Continually deepening effects

The original concept of the builders of the first railways (in other words, those who harnessed the new technology) was quite simple, namely to use the mechanically hauled carriages, running on rails to enhance their effectiveness, to link together related towns. The first railways were not built as networks, but on a point-to-point basis. However, within only a very few decades a network had developed, but also a whole range of ancillary yet essential supporting measures. The railways quickly espoused new concepts and new technologies, and with this the necessary mass education. The result was significant social change, initially at local and regional level,[2] and the invention of essential supporting aspects.

Safety and science

As soon as railway traffic and travel speeds built up, increasing safety measures were needed. This led to the development of mechanical signalling and many related mechanised safety devices, including some very sophisticated and complex adaptations of levers, pulleys and other concurrently state-of-the-art technology. Then, in a dress rehearsal for telematics, the railways quickly espoused and developed the newly invented electronic telegraph in order to achieve near-instant communication.

Time travellers

Another necessity of the railways was the development of standard time. Hitherto each major town or village could work according to local time, but it was not possible to run railways on this basis. Thus nationally based standards introduced as a necessity for the railways replaced precise solar accuracy for the community, in order to support technological gain.

Essential standards and operational rules

Although time could no longer stand still, this was only one of the essential new standards. Within individual companies there had to be prescribed ways of undertaking the multiplicity of operational tasks – both routine and to cope with adverse incidents – as well as a new focus on standard accounting and reporting. However, as railways linked up, inter-operability made standards even more essential if trains were to move safely between operators. Unfortunately, on occasion it took either an avoidable accident or a commercial crisis for standards to be established and agreed upon.

Staff education and training

The operation of railways required widespread use of the written word – for operating requirements, instruction manuals and the burgeoning working-rule books. However, the railways were recruiting artisan staff who had never before needed high levels of literacy. Thus the railways embarked on a major and underappreciated mass-education movement to bring literacy skills to its work-force. This educational infrastructure then also provided the vehicle by which technical training could be applied.

Social engineering

The railways were also the creators of other social engineering. Built to link towns, they soon themselves created towns. New railway industries emerged, and as a consequence major new towns sprang up directly to service the railways, and often largely funded by them. Thus housing standards, economic development and social restructuring were all inevitable consequences of what originally had seemed to be merely a new method of mechanical transport.

Major indirect effects

Once the transport of people, and also of goods, became economical, industrial specialisation was facilitated. Efficient industrial locations expanded while the inefficient or low-quality providers were forced out. Not only industry was affected as, for instance, the railway also benefited remote dairy farmers who now had the means of conveying their highly perishable product to the new towns and cities.

None of the initial railway pioneers – from those who put the first engine on to wheels through to the builders of the first inter-town railways – could have envisaged the degree to which the new transport medium would trigger major social and industrial upheaval. However, with the wisdom of hindsight, it appears inevitable.

Health telematics forewarned

And what of health telematics? The preceding chapters of this book have illustrated what telematics in health (and in particular telemedicine) can achieve, but also that there is a commensurate need for a controlled harnessing of this new technology if potentially serious problems are to be avoided. The analogy between the railways taking the time out of distance and telemedicine taking the distance out of healthcare is a clear one as the consequent requirements are virtually the same, namely new safety procedures, new applied technology, more sharply defined time, agreed technical standards, formal operational procedures, education of staff in new skills as a precursor to specific technical training, and the generation of specialist enabling activity which is supportive of the core purpose.

Fibre optics and chips – the new social sustenance

Furthermore, in the new 'information society', other telematics applications such as on-line databases and the Internet will radically alter both practice and public perception by bringing remote information immediately to hand. It is already commonplace for the public to visit their doctor with a self-analysis of their symptoms and a suggestion as to what their treatment might be. Similarly, patients are not afraid to use the Internet to find out more about the treatment which has been prescribed, and to challenge their clinician with regard to the choice of treatment or medication (sometimes with good reason). They also have the technical potential to seek for themselves a second opinion, quite possibly from another country. Yet there are considerable dangers in this empowerment, ranging from the threat to the doctor–patient relationship, through to the risks of the public not fully appreciating the nature and cultural setting of the information which they are accessing, as well as the lack of effective controls of authenticity and quality of that information. Members of society have every right to seek to enjoy the new enablement, but society also needs to ensure that they are protected against new adverse effects, and the global health sector needs to balance enablement with the potential for destabilisation.

Forward in safety

The problem is not that telecommunications and telematics should be applied to healthcare, but that this should be happening without any adequate mechanisms for consumer protection against the threats identified in the first and several subsequent chapters, namely direct threats to individual well-being, and indirect threats through changes to the local healthcare sector. Although the purpose of any individual telematic or telemedicine application is to improve treatment opportunities, the growing effects are unlikely to be that simple. Indeed, it is questionable whether there is any reason to think that telematics and telemedicine will have any less radical societal and professional effects in their first 30 and 50 years than the railways did. Here, too, the effects can also be seen as generically similar, namely a general societal and commercial desire to harness the new technology, followed inexorably by legitimate anxieties about new safety hazards, and then by social and economic re-engineering as activities move to locations where they are most efficient or cost-effective.

The additional new telematics effects

As healthcare moves into the new millennium, telematics and telemedicine will inevitably move to take the distance out of clinical practice. However, as telematics and telemedicine are at the dawning of their own new era, not only will the electronic equivalent of the transport automation paradigm shift come into play, but a whole new generation of effects will also have to be taken into account. It is too early to define these in detail, and thus it is difficult to be precise about the consequences or the definition of related solutions, but it is vitally important to attempt to highlight the issues. All of those involved – health professionals, the clinical professions corporately, governments and the international community – will have to be prepared. Given our improved awareness of the potential of new technology to induce behavioural and organisational change, it is no longer acceptable to sit back and wait until adverse effects become obvious. The following section outlines important issues.

Knowledge

To date, clinical practice has been restricted by the amount of knowledge which could be obtained from any health professional. However, telematics solutions such as electronic publishing and knowledge bases, particularly through the Internet, have changed the situation from data shortage to data swamping. Thus, without relevant filtering and interpretation mechanisms, the clinician's knowledge may be no more perfect than before, if knowledge is considered to be that information which is readily available related to the current activity in an accurate and appropriate way. New approaches to seeking and appraising information are becoming necessary as well as appropriate.[3]

Clinical skills

The clinical community always tends to learn too late that new technologies and techniques need new skills. The previous chapter explained how minimally invasive surgical procedures were shown early on to have potential advantages to patients compared to open abdominal procedures in certain circumstances, but in fact the hand–eye co-ordination skills required when working remotely through fibre-optic scopes and television monitors are radically different to the skills utilised through direct sight and touch in an open incision, resulting in problems with inadequately prepared replicative implementation.[4] In telemedicine specifically, the skills of remote visualisation, history-taking, examination or operation will be different again (as well as being dependent on the skills of a remote operative who may be totally unknown). Teleconsultation needs different communication and observation skills from face-to-face contact, as well as compensation for loss of direct use of important senses, particularly touch and smell. Even the wider techniques of telematics, with the ability to present vast amounts of information either from the electronic patient record or from reference knowledge bases, will require the health professional to have very different information-searching and assimilation skills compared to those required for the reading of paper records and references.[3]

Trusting what's on screen

As indicated in the previous chapter, in the manuscript era the clinical record could generally be trusted within its use environment because it was built up by members of a distinct and mutually known clinical community, with the authorship of each entry being identified. The electronic record, by contrast, will be built up by a virtual community of clinicians who may be spread over a much wider area, and who are not necessarily known to one another. Electronic techniques such as automatic identification of authorship and electronic signatures, when adequately employed, can provide attribution for the purposes of audit and accountability, but this will give little guidance with regard to the clinical decision-making processes and criteria of a remote clinical colleague.

Similarly, whereas textbooks and journals carry their own explanation of refereeing and other authentication processes, electronic reference sources intrinsically do not have the same essential mechanisms for authentication or control of accuracy, and some sites are intentionally misleading or misinformed whilst at the same time giving every visual appearance of being highly professional. There is recognition of the need for validation techniques by the reader,[5] whilst other initiatives seek to address the development of authentication mechanisms.[6-9]

Conflicts within public expectation

The public in general is well aware of the potential of telematics and telemedicine, from sources which range from their own mere anecdotal understanding of the Internet through to popular television programmes which tend to report the latest technological developments out of context of the new issues and needs for controls which they raise. Consequently, the public tends to push for rapid harnessing of the latest technology, yet on the other hand increasing consumer awareness results in expectations of safety features and controls which are an amalgam of the best of the pre-electronic era with the best that electronic mechanisms can offer.

Health professional dilemmas

For the individual health professional, telematics systems create opportunities both for personal updating, and for drawing upon reference sources. At the same time, the unstructured and frequently unverified nature of the sources of available information can be daunting as well as time-consuming. Not to use the most recent evidence available may be deemed unethical, but at the same time using guidelines which turn out to be inappropriate can put the clinician in a position of personal liability. A balance needs to be struck in order to enable structured assistance in appraising evidence, without controlling freedom of innovation or thought.

Validating remote and informal learning

A potential benefit of teleconferencing and similar applications is structured remote learning, assisting clinicians who would have difficulty in travelling to learning environments. In addition, as indicated in earlier chapters, one frequently mentioned spin-off from telemedicine in particular is that remote clinicians can learn clinical skills through the process of facilitating expert telematic support for the patient. However, the preceding chapter has emphasised the importance of such remote learning being validated in order to avoid undermining of the recent important advances in optimising skills transfer and application of knowledge.

Geographical specialisation

The one aspect on which it might initially be assumed that telematics would not have an untoward effect is that of geographic centralisation (indeed, commercial call centres are bringing employment to remote locations). However, as areas of clinical expertise grow, so geographical clusters may be expected to develop. Already, even ahead of the widespread effects of telemedicine, strategies for regionalisation of cancer services exist in order to maximise skills,[10] so communities of clinical expertise are perforce likely to build up. With the advent of telemedicine, there are already trends in some parts of the world to harness external skills. For instance, emergent-economy countries expanding the availability of computerised axial tomography (CAT) scanning may find it more effective to contract the specialised reading of the scans out to overseas centres which already have that expertise. As a result of this approach a clinical equivalence to geographical clustering of computing expertise may develop, as expert centres grow in both demand and clinical aura. Thus Telemed City may emerge as a clinical equivalent to Silicon Valley and Silicon Glen. Whether this global equivalent of Teaching Hospital Syndrome, drawing together a community of clinicians with advanced and ever-deepening special skills, will be progressive or regressive is an interesting subject for consideration.

Towards the global clinic

Thus the impending arrival of the global clinic – an entirely feasible concept based on existing health telematics and telemedicine technology – raises issues which neither society, governments, nor global organisations have begun to address in any depth.

Public expectations

In an era of global communications, the public expectation is of the best available treatment, and its views are no longer constrained by local experience. The resident of the Caribbean island where advanced tertiary care is not practicable is nevertheless likely to expect the same standard of treatment for a cardiac or orthopaedic condition which they see on the American television channels which they view nightly. Even if, over a period of time, telemedicine links were to be installed to enable consultation and even remote treatment to be carried out, this would still not solve the problem of the correct therapeutic and rehabilitation environment. Expectation is raised, but virtual reality is not the same as realism.

Consumer choice

A further and generally justifiable consumer attitude is that of choice – choice of type of treatment, and choice of individual practitioner. By taking the distance out of clinical care, the consumer can in theory now consult any other medicine practitioner, or a particular well-renowned specialist, without travelling. The logistics of how the appropriate practitioner is selected on the basis of evidence rather than advertising, and how the patient can make an appropriately informed decision, have yet to be resolved.

Clinical training and burnout

Clinical education, particularly of doctors, is recognised as being a process which takes many years. With the traditional hierarchy of primary, secondary and tertiary care, with their concentric catchment areas and access being controlled by a broad balance between need and the barrier of travel, a stable structure has been established. However, this has also been based on the unavoidable necessity for patients to be seen in part by clinicians who are learning as they practise. The line between being treated by a student (which requires informed consent) and being treated by a health professional undergoing supervised learning (which is seldom explained) is an indistinct one. With the availability of teleconsultation, patients and indeed their primary care physicians can instantly pass through the barrier of distance to consult directly with the best available expert in the country, or indeed in another country. If (or when) this facility becomes widespread, it will remove the foundations from the clinical learning structure. At the same time, it would not be long before the best available experts suffered from burnout, exacerbated after a short while by shrinking of the layer of intermediate skill levels.

The new ethical challenges

Thus telemedicine and telematics bring new ethical challenges. Up until now distance and the practicality of handling large volumes of paper-based information have provided *de facto* filtering mechanisms. The test of appropriate clinical behaviour – that of 'reasonableness' – has been based upon establishment of referral hierarchies and recognised personal learning mechanisms. Telematics challenges and potentially demolishes both of these. Total freedom to travel in virtual time to any clinician on the globe, or to expect the local health professional to identify and implement the best information appropriate to the clinical circumstances, are each in effect an expectation of the impossible. Yet the alternative is to create new filtering mechanisms, taking away the new opportunities which the electronic era produced.

It has taken 2000 years to produce working and clearly understood mechanisms to put into daily use the key ethical principles of current clinical practice. The four principles of Beauchamp and Childress[11] (namely respecting autonomy, beneficence, non-maleficence and justice), for instance, draw heavily on the enduring principles of Hippocrates.[12] The societal challenge is to provide the mechanisms for the electronic era to ensure that these same principles apply reliably in the global clinic – but to do so in 2000 days.

Local and national legislation

In many cases national legislation lags far behind the electronic era – for instance, requiring medical documents to be in a paper-based format, and with a personal signature. Regulation to practise and principles of legal accountability are nationally (or even state) based. These requirements continue to dominate in an era when land boundaries no longer have the same practical meaning, and indeed it may not be obvious in which country an information source or a practitioner is located.

However, against this background the mechanisms do not yet exist for the agreement or enactment of international legislation to regulate global clinical practice.

The challenge of global governance

The underlying and ultimate challenge for the safe application of health telematics is that there is no global government, but telematics raises a host of global issues. On the one hand, health telematics is very progressive, with its power to bring advanced scientific knowledge to economically poor settings, and to enable the citizen to circumvent governmental or profession-imposed curbs to freedom of information or advice. On the other hand, however, citizens and communities deserve protection against adverse effects, whether these are simply the perverse effects of real-time global consulting or more sinister manipulative effects. At the very least, there is a global moral duty to anticipate, identify and advise, so that all concerned take informed action.

The key societal protection issues

A few key issues can be identified, and these are summarised below.

- *Trust* – is the service, or the practitioner, trustworthy?
- *Relevance* – is the service or advice relevant to the remote setting from which they are being consulted, and what is the science base?
- *Quality* – is the quality the best reasonable?
- *Enablement* – does the service benefit the person or user professional, without them becoming inadvertently locked into ongoing dependence?
- *Security* – is confidentiality assured, and are personal data transmitted and stored securely?
- *Liability* – are there means of ensuring liability, accountability and redress in the event of an adverse effect?

No one currently using any telematic service across a national boundary can be assured of these points. Telematics could be summed up as '*trans* everything, *sans* anything'. The scope of individual governments is limited, and even then may not always be totally appropriate. For instance, some governments support their clinicians in practising remotely as being good for trade and prestige, but prohibit other nationals from providing advice to their own citizens – a paternalistic position whose absolutism may not be justified.

Global spheres of interest

Global action is not easy, but health telematics is not the first in the field – for instance, air traffic control is successfully controlled globally. However, air travel grew from a new beginning, and was also initially the province of the influential, so new organisations and agreements were easier. Health and healthcare have long traditions in most societies, and affect every citizen in an intimate way, so their very importance creates the international complexity. It is therefore more appropriate to focus on the constituent spheres of interest, and to relate these to existing global bodies.

Health

The World Health Organization (WHO) concentrates on global health issues, not specific care delivery. However, the issues of the global clinic should be seen as global issues, and their effects must be viewed as affecting health and healthcare globally, every bit as much as a new pandemic. The WHO must be the appropriate lead body, but with telematics as a new dimension of the global health agenda.

Trade

Telemedicine, telematic services and commercial aspects of Internet medicine are clearly trade activities, and of a particularly significant nature. This is the competent province of the World Trade Organization (WTO), which applies a strong social dimensional interest to its work in other domains. If the WTO can take action to ensure that the trading power of large economies does not have an adverse effect on the economic well-being of farmers in fragile economies, it must surely have a moral duty to exert the same influence on global health trade to protect the health of populations.

Standards

An important aspect of telematics and telemedicine is ensuring standards, ranging from telecommunications standards through to terms, language and prescribing. Here the International Standards Organization (ISO) has the lead role, linking the various regional bodies that are involved in health standards.

International priorities?

Given the challenges, what then are the priorities for progress and protection? With the wisdom of the twentieth century moving into the twenty-first century, it is surely unethical to adopt the approach which was inevitable with the innocence of railway development, namely to assume that there will be no changes and then to be proved wrong. We know that there will be effects, both locally and globally. At a global level, these must be distilled to the essential. Five priorities can be postulated, and these are listed below.

Visioning

Drawing on the initial work of the WHO in health telematics,[13] and using futures analysis techniques as developed not least by the Dutch,[14] and most recently by the Nuffield Trust in conjunction with the Judge Institute of Management in Cambridge,[15] regular structured brainstorming could and should usefully build on the modest start made by the foundational London workshop (see Chapter 1) and this volume, to identify potential new issues. Fore-warned is forearmed.

Observational monitoring

Watching what happens is the best way to identify trends in order to analyse the benefits and adverse effects, and from there to strengthen the benefits, and control with minimal intervention the potential adverse effects. A non-interventionist observatory approach has major attractions, of which examples already exist, including the Danish Health Telematics Observatory.[16]

Labelling

The key issue for both individuals and professionals is informed action. If they are to use Internet sites, telematics sites or telemedicine services, they need to know the status of these. Labelling, in the sense of displaying standardised information to enable them to make informed choices, is an effective means of empowering the public and user professionals without restricting freedom of choice. Possible items that would appear on standard labels for telemedicine services or Internet sites might include confirmation of the country (or countries) in which the lead clinician is registered, the country of location, the published scientific basis of the site/treatment, the insurance backing and the professional supervisory body.

Standard setting

Many standards have already been adopted successfully in telecommunications, and also in many area of health informatics. The need is to draw these together into international health telematics standards, not least in terminology, to ensure non-ambiguity and reliability. For instance, much more needs to be done to standardise the conveyance of meaning, which is fundamental to health.[17]

Researching the unknown

Health telematics in the global setting raises new questions which can only be addressed very imperfectly at present, as they seek to anticipate global behaviour and effects. These include whether bringing leading-edge techniques and knowledge to poorer economies will improve care or whether it will lead to impossible further expectations. Will cost containment result or will poor countries be forced into higher and unsustainable spending in order to meet demands to use the newly available knowledge? Will quality of care improve or will remote application be inaccurate and thus harmful? Will local skills development be enhanced or will ambitious practitioners migrate to the new global centres? Can countries develop comprehensive plans to exploit the new opportunities in locally appropriate and sustainable ways? Will local centres of expertise feel empowered or will there be a continuing regressive tendency to focus on the research, and thus the science base, of the most affluent countries? Will local staffing and training structures be strengthened or destabilised?

An international health telematics body

Clinical governance is increasingly recognised as the essential means of corporate responsibility in the health sector.[18] It involves applying a professional social responsibility and accountability framework, without inhibiting acceptable clinical freedom and the professional–patient relationship. Whilst such structures are local and national matters for traditional care, telematics passes through those controls.[19] A new global governance paradigm is needed for the global clinic.

However, in a global society which is wary both of any form of regulation and of the costs of supporting global bodies, are such global clinical governance safeguards likely to be achieved? There is an encouraging model. Food is vitally important to global society, and a successful organisation has been established jointly between the World Health Organization and the Food and Agriculture Organization (FAO), initially to develop labelling standards for food and now also covering wider standards. This is the Codex Alimentarius Commission, whose prime objective is to publish the *Codex Alimentarius*,[20] a standard vocabulary of terminology which is used internationally. The decision of governments to use, and to underpin by legislation, the *Codex* is voluntary but the majority do so. The Commission also holds regular summit meetings on key topics and produces resultant accepted published standards, such as those on food hygiene and on maximum residue levels for pesticides.[21]

Similarly, civil aviation activity is regulated by a global body, namely the International Civil Aviation Authority, a United Nations agency based in Montreal and mandated by the Chicago Convention of 1944. This develops and applies evidence-based standards for equipment, processes, services and expert personnel (for the latter, with regard to both their skills and their health), in order to protect the public in a context of free trade and global freedom. Some 185 countries have signed up to this convention, and are thereby required to give mutual enforcement and recognition of the agreed standards. Ironically for a consumer safety-focused process, the one activity which falls outside this regulatory framework is the newly emergent provision of tele-medicine services to airlines to provide assistance with medical emergencies in flight.

Given the key importance of healthcare, and the inevitable inter-nationalisation of health telematics, it seems not unreasonable to see the *Codex Alimentarius* and civil aviation as potential models for a health informatics body – enabling, informed, evidence based and authoritative, and thereby providing a basis for unilateral but compatible government action. To an even greater extent than with food, but similar to aviation, the solution has to be globally owned, as the protection of the user citizen must involve standards implemented by the countries that offer the services. A Health Telematics Commission, co-sponsored by the key world bodies, could initiate work in the areas identified earlier and develop standards frameworks which could be endorsed by member governments, as well as sponsoring global outcomes research.

Conclusion

Clearly the global clinic already exists, and is open for business as we move into the twenty-first century. Increasing numbers will be tempted to visit it, for a variety of good and practical reasons. However, they will not be entering the equivalent of a conventional clinic based elsewhere, with the underlying assumption of existing terrestrial rules. Instead, they will be moving into an environment of which there is only limited practical understanding, and that has very little underpinning by quality control and legislative frameworks.

At the same time, however, to inhibit use of new opportunities would itself be unethically restricting. As with the mechanisation of transport, both the overall objective and the individual elements are ethically justified and intended for the advancement of human health. As with the development of the railways, individual components can be planned and controlled but drawing on the same past experience, it can safely be predicted that unforeseen issues will arise, as well as all of the new challenges which have been briefly outlined here. In short, the global clinic needs global clinical governance processes – which in turn should be enabling and not restricting. Telematics and telemedicine are thus issues for which global organisational frameworks are needed. The creation of these is the next fundamental challenge in the furtherance of health telematics as we move into the twenty-first century.

References

1 McLuhan M (1964) *Understanding Media: the extensions of man.* Routledge, London.
2 Sylvester D (1971) *A History of Cheshire.* Darwen Finlayson, Henley-on-Thames.
3 Roberts R (1999) *Information for Evidence-based Care.* Radcliffe Medical Press, Oxford.
4 Banta HD, Schersten T and Jonsson E (1993) Implications of minimally invasive therapy. In: HD Banta (ed) *Minimally Invasive Therapy in Five European Countries: diffusion, effectiveness, and cost-effectiveness.* Health Policy Monographs Vol. 3. Elsevier, Amsterdam.
5 Tyrrell S (1999) *Using the Internet in Healthcare.* Radcliffe Medical Press, Oxford.
6 www.hon.ch
7 www.multimedica.com/TEAC
8 Eysenbach G and Diepgen TL (1998) Towards quality management of medical information on the Internet: evaluation, labelling, and filtering of information. *BMJ.* **317**: 1496–502.
9 Kim P, Eng TR, Deering MJ and Maxfield A (1999) Published criteria for evaluating health related web sites: review. *BMJ.* **318**: 647–9.
10 Expert Advisory Group on Cancer to the Chief Medical Officers of England and Wales

(1995) *A Policy Framework for Commissioning Cancer Services: guidance to purchasers and providers of cancer services.* Department of Health, London.

11 Beauchamp TL and Childress JF (1989) *Principles of Biomedical Ethics.* Oxford University Press, New York.

12 www.emory.edu/WHSC/MED/HTN/~achung/hippocrates.html

13 World Health Organization (1998) *A Health Telematics Policy.* World Health Organization, Geneva.

14 Steering Committee on Future Health Scenarios (1988) *Anticipating and Assessing Health Care Technology* (series). Kluwer, Dordrecht.

15 Dargie C (1999) *Policy Futures for UK Health: Pathfinder – a consultation document.* The Nuffield Trust, London.

16 Vingtoft S, Koldsø N, Lippert S *et al.* (1998) *A National Project for Promoting EPR Development: the Danish HOP project.* In: PW Moorman, J van der Lei and MA Musen (eds) *Proceedings of the IMIA Working Group 17,* 8–10 October 1998 (*EPRiMP: The International Working Conference on Electronic Patients Records in Medical Practice,* Department of Medical Records, Erasmus University, Rotterdam).

17 Sheridan AJ, Rigby MJ and Draper RJ (1999) From bridges to super-highways: transmitting meaning within and between professions, and across time and space. In: P Kokol, B Zupan, J Stare, M Premik and R Engelbrecht (eds) *Medical Informatics Europe 1999.* IOS Press, Amsterdam.

18 Secretary of State for Health (1997) *The New NHS: modern, dependable.* The Stationery Office, London.

19 Rigby M (1999) The management and policy challenges of the globalisation effect of informatics and telemedicine. *Health Policy.* **46**: 97–103.

20 Joint FAO/WHO Food Standards Programme (1998) *Codex Alimentarius: food labelling complete texts.* Food and Agriculture Organization of the United Nations and World Health Organization, Rome.

21 www.fao.org

Index

Access to Health Records Act, 1990 111, 122

accreditation, medical practitioners 13, 110, 132

advertising
  on Internet 56–7
  telemedicine services 110

air traffic control 201

Akershus County 77–8

ambulance booking service 144, 145

American College of Cardiology 132

appointments, accelerated, reserved 73

armed forces 23

Association of Community Health Councils for England and Wales (ACHCEW) 152–3, 158

attitudes
  measuring 69
  to computers 54
  to telematics 43, 44, 47–8, 87–8, 183

audiotape recordings 111

audit, clinical 136–7, 158

autonomy
  in nurse-led telemedicine 97–8
  respect for 159, 160–5

Belfast, nurse-led telemedicine 92, 156

bias, data-collection 69

billing mechanisms 179

blinding of studies, difficulty in 67–8

booking systems 179

British Telecom (BT) 155

burn-out 174, 198–9

Caldicott Report (1997) 124, 133, 157

cameras, document 35, 36

cancer services 155, 197

CANDELA project 53

Central Middlesex Hospital, London 92

Centre for Health Informatics, Swansea 86

centres of excellence 174, 197

Ceredigion Community Health Council 154, 155, 157

champions, telemedicine 18

change
  culture 89, 144
  managing 147
  obstacles to 48
  planning for 139–40, 148, 184
  rapidity of 66, 146
  in working methods 45, 47–8, 174–5

CHCs see Community Health Councils

check-list effect 70, 72

Civil Jurisdiction and Judgements Acts, 1982, 1991 109

Cleveland Wide Area Network (CWAN) Project 143–8
  aims 145
  constraints encountered 145–7
  education and development 148
  managing change 147

clinical chemistry, Internet-based knowledge base 51, 52–4

clinical culture, change in 89, 174–5

clinical governance 130, 137
  definition 137
  global 203–4, 205

clinical practice, effects on 173–8, 194–5

clinical process, quality of 181

clinical skills
  destabilisation of development 174
  informal learning 175–6
  need for new 195

clinical support systems
  barriers to integrated 105–6
  introduction of system in Teesside 143–8

clinicians see doctors

cluster randomised trials 71, 87–8

Cochrane Effective Practice and Organisational Change Group 69

Code of Conduct, for Internet-based services 58–9

Codex Alimentarius Commission 204